美食城堡的安全保卫战
A War to Protect Food Safety

沈建忠　张嵘　主编

知识产权出版社
全国百佳图书出版单位

图书在版编目（CIP）数据

美食城堡的安全保卫战 / 沈建忠，张嵘主编 . —北京：知识产权出版社，2018.4
ISBN 978-7-5130-5466-9

Ⅰ . ①美… Ⅱ . ①沈… ②张… Ⅲ . ①食品安全—青少年读物 Ⅳ . ① TS201.6-49

中国版本图书馆 CIP 数据核字（2018）第 050270 号

内容提要

本书为专门介绍食品安全小常识的科普读物，针对当前食品安全中普遍存在的隐患，以及广大青少年最为关注的一些食品安全问题，简明扼要地讲解了日常生活中各种常见食品的鉴别、选购方法、食用注意事项及其他相关基本知识。

责任编辑：于晓菲　　　　　责任出版：孙婷婷

美食城堡的安全保卫战
MEISHI CHENGBAO DE ANQUAN BAOWEIZHAN

沈建忠　张　嵘　主　编

出版发行：	知识产权出版社有限责任公司	网　址：	http://www.ipph.cn	
电　话：	010-82004826		http://www.laichushu.com	
社　址：	北京市海淀区气象路 50 号院	邮　编：	100081	
责编电话：	010-82000860 转 8363	责编邮箱：	259451929@qq.com	
发行电话：	010-82000860 转 8101	发行传真：	010-82000893	
印　刷：	三河市国英印务有限公司	经　销：	各大网上书店、新华书店及相关专业书店	
开　本：	720mm×1000mm　1/16	印　张：	11	
版　次：	2018 年 4 月第 1 版	印　次：	2018 年 4 月第 1 次印刷	
字　数：	168 千字	定　价：	58.00 元	
ISBN 978-7-5130-5466-9				

出版权专有　　侵权必究

如有印装质量问题，本社负责调换。

编写委员会

主 编：

沈建忠　张　嵘

Chief Editors:

Jianzhong Shen　Rong Zhang

编辑和插画：

赵贺敏　张仲阳　张美燕　王慧敏
黄子衔　王汉宇　李佳萍　李金樾

Editors and Illustrators:

Hemin Zhao　Zhongyang Zhang　Meiyan Zhang
Huimin Wang　Zixian Huang　Hanyu Wang
Jiaping Li　Jinyue Li

翻译：

乔伊·坦　沈张奇　汪洋　王战辉

Translaters:

Joy Tan　Zhangqi Shen　Yang Wang　Zhanhui Wang

前 言

自古以来，民以食为天。食物作为人类生存必需品，其重要性不言而喻，而食品安全更是关乎人类健康和社会稳定的民生问题。

近年来，国内外都有食品安全事故的报导，这些个别的、却触目惊心的事件，不仅危害人体健康，更是严重干扰了社会经济秩序，遗患无穷。中国政府高度重视食品安全问题，将食品安全列为《国家中长期科学和技术发展规划纲要（2006—2020）》（以下简称《纲要》）公共安全领域的优先主题。国务院成立了高层次国家食品安全委员会及其办公室，先后颁布了《中华人民共和国农产品质量安全法》《中华人民共和国食品安全法》和《中华人民共和国食品安全法实施条例》，严格规范安全食品生产销售。

少年儿童是祖国的未来和希望，却是食品安全问题的最大受害者。为了保证少年儿童的健康成长，普及食品安全知识，杜绝不安全食品，进行食品安全普及教育势在必行！

中国农业大学和浙江大学的师生们联手一起奉上《美食城堡的安全

美食城堡的安全保卫战
A War to Protect Food Safety

保卫战》，这本中英文双语科普读物沿袭了科普读物《细菌与抗生素之战（一场肉眼看不见的战争）》的风格，用活泼生动的语言，以故事的形式讲述了五个卡通形象身边的食品安全事件，内容贴近生活，趣味盎然，同时又宣传了食品安全知识，会使小读者受益匪浅！

Preface

Sustenance is of utmost importance to people. Without a doubt, food is fundamental to survival, and food safety can significantly impact a society's social stability and the public health of its people.

In recent years, there were numerous food safety accidents, these cases have been extremely problematic, and it went on harming the people's health as well as the country's social stability. The Chinese Communist Party and government attach great importance to food safety, listing it as a thematic priority in the public safety field of China's National Policy for Medium and Long-term Scientific Development (policy). The State Council of China formed a high-level food safety committee and issued a Code of Primary Products Quality Security, Food Safety Law and other regulations to standardize food production and distribution in the country.

Children and adolescents are the future of our country. However, they have become the biggest victim of food safety problems. Therefore, in order to promote the health and growth of our children and adolescents, to spread knowledge on food safety and eliminating unsafe foods are imperative steps to secure the society's future.

美食城堡的安全保卫战
A War to Protect Food Safety

To further spread food safety knowledge among the next generation, and in order to increase their awareness and self-defense capability, China Agricultural University and Zhejiang University jointly wrote a Chinese-English science reading named A War to Protect Food Safety following the success of the previously published reading, Invisible War: A Battle Between Bacteria and Antibiotics. By telling stories about starring five cartoon characters, this science reading propagates food safety awareness among children, and helps them better understand scientific knowledge in a simplified and fun way.

目录

"五毛零食"大爆炸
The explosion of the "50 cents treat" / 1

"毒馅"小作坊
The Mill with "Poisonous Fillings" / 12

爱"洗澡"的草莓
The Strawberry Who Loves to Bathe / 20

被密封的水果宝宝
The "Contained" Fruit Babies / 25

奔跑吧,外卖君
Go Deliveryman, Go! / 33

变身芽芽的土豆君
The Potato with Tails / 43

常吃爆米花,肥胖和你不分家
Too much Popcorn Makes you gain a lot of weight / 49

美食城堡的安全保卫战
A War to Protect Food Safety

腐臭豆腐和致癌咸菜的pk赛
Stinky Tofu VS. Pickled Vegetables / 57

翻滚的麻辣烫君
Rolling Malatang / 65

果汁色彩斑斓的秘密
Colorful Secret of Fruit Juice / 72

黑暗料理风波
The Mysterious Dish / 78

黑心小作坊
The Dishonest Shop / 84

胡萝卜让你眼睛美又亮
The Beautifying Carrot / 92

化妆的鸡蛋
An Egg with Makeup / 99

皮蛋皮蛋是坏蛋
Preserved Egg is a Bad Egg / 107

烧烤架上的大坏蛋
The "Bad" on the Grill / 117

生的海产品，我们不约

Say No to Raw Seafood　/ 125

香蕉宝宝的爱与恨

Love and Hate of Banana Baby / 135

香甜诱惑

Temptation of Sweet / 140

虫眼 + 果蔬 = 绿色？

Bug holes+Vegetables=Green? / 145

炸鸡的正确打开方式

Eat Fried Chicken with Right Way / 154

书中人物介绍

Character Introduction

小微博士 Dr. Micro

特性：上懂天文，下晓地理

Extremely intelligent and knowledgable

松仔 Squirrel

外号：智慧星 a.k.a. Smarty

特性：聪明伶俐，善于思考

Smart

雨燕 Emily

外号：及时雨 a.k.a. Best Partner

特性：飞行速度快，乐于助人，松仔的得力小助手

She can fly really fast, and loves to help others. She is Squirell's best assistance.

胖胖熊 Little Bear

外号：淘气鬼 a.k.a. Trouble Maker

特性：最大的爱好就是吃吃吃，一个十足的淘气鬼

He just loves to eat and play!

狐大哥 Mr. Fox

外号：老江湖 a.k.a. Old Timer

特性：无所事事，不爱学习

He likes to have fun, but tries to avoid studying at all.

美食城堡的安全保卫战
A War to Protect Food Safety

"五毛零食"大爆炸

The explosion of the "50 cents treat"

下课的铃声"叮叮当当"地回荡在美食城堡小学的每一个角落，胖胖熊和一群小伙伴井然有序地做着值日的最后一项打扫工作。

Class dismissed and the recess bell resonated around every corner of the Castle Elementary School. Little Bear and his friends were on cleaning duty that day, and they just tidied up the last bit of the classroom.

"终于全部搞定啦！"，胖胖熊放下扫把，顺势伸了个懒腰，露出圆滚滚的肚子，"辛苦你啦，带你去吃包辣条减减压"，他拍了拍自己的肚子，喃喃自语道。

"We finally finished!" said Little Bear, he put down the broomstick and stretched his body until his little belly turned up. "A job well done, time to treat you to some delicious rice-sticks!" Little Bear patted and said to his belly, he was ready to head out for a delicious treat.

一踏出校门，胖胖熊就径直跑向了马路对面的一家小商店，不一会儿便抱着一堆带有花花绿绿的

As soon as he stepped foot out of the school gate, he ran straight to the convenience store across the street. In

美食城堡的安全保卫战
A War to Protect Food Safety

包装袋走了出来。只见胖胖熊抑制不住满脸的喜悦之情,马上撕开一包,美滋滋地享用了起来。此时,小微博士开车路过,恰巧看见正在等校车的胖胖熊。

"胖胖熊,你跟老师说一声,我载你回家。"小微博士将车停在一旁,摇下车窗冲着胖胖熊招手。

"好嘞!"胖胖熊开心地直奔过来,一股脑儿地将书包和手中的零食抛在后驾驶座上,便跑去向老师请示了。

顷刻,车内便充满了一股"香浓"的味道,小微博士转头望向后

no time, Little Bear walked out of the shop with many colorful bags full of all sorts of snacks. Little Bear had a big smile on his face, he couldn't wait to open the bags of deliciousness. As he began to eat, Dr. Micro happened to drive by, and she saw Little Bear beside the street with his snacks.

"Little Bear! I can give you a ride home, just let your teacher know first!" Dr. Micro parked his car on the side of the street, rolled down his window and waved at Little Bear.

"That would be wonderful!" said Little Bear with excitement. He opened the car door, threw his backpack and all the snacks onto the passenger seat, and quickly ran to tell his teacher.

Immediately, the car was filled with a rich and pungent smell. Dr. Micro

美食城堡的安全保卫战
A War to Protect Food Safety

座。只见那堆零食中以辣条、辣片等调味制品为主，还有几包果脯、膨化食品和袋装饮料。它们的名字五花八门，包装花花绿绿的，口味以甜和辣为主。小微博士随手捡起几包辣条，光凭外包装的油腻触感，就让她不禁对食品的安全性感到担忧。小微博士发现，这些辣条上都标有"QS"（企业食品生产许可）标识，但仔细一看，这些零食里可添加了不少食品添加剂，甚至在某些零食中添加剂多达几十种。

小微博士在认真查看配料表后发现，这些辣条都含有3大类食品添加剂。

turned to look at the back seat of the car, and she saw several bags of spicy rice-sticks, chips, and other snacks made with plenty of flavoring. There were dried fruits, puffed snacks, and bagged beverages. The snacks had such colorful packaging and names of all sorts, and they were either spicy or sweet in taste. Dr. Micro picked up a couple of them, the oily packaging was enough for her to worry about whether the snacks were safe to eat. Dr. Micro realized that, although the packaging all had a printed "QS" certification on it, the snacks actually contained a lot of additives and flavoring, some snacks even contained more than a few dozen varieties of additives.

After Dr. Micro carefully studied the ingredients printed on the back of the packages, she realized that there

美食城堡的安全保卫战
A War to Protect Food Safety

常见的甜味剂类包括阿斯巴甜（含苯丙氨酸）、甜蜜素、三氯蔗糖、安赛蜜、纽甜。阿斯巴甜最初是在合成促胃液分泌激素时偶然发现的具有甜味的物质，它比蔗糖甜约200倍，在人体内会代谢产生苯丙氨酸，苯丙酮尿症患者不能食用，所以现规定含有阿斯巴甜成分的食品必须标明含有苯丙氨酸；甜蜜素摄入过量会对人体的肝脏代谢和神经系统造成危害；三氯蔗糖是由蔗糖制取的，目前没有充分证据证明其存在毒性，是一种比较理想的甜味剂，但其制取难度大，价格高。

were three major food additives in the spicy rice-sticks.

The most common types of sweet additive include APM (aspartame), sodium cyclamate, TGS(trichlorosucrose), acesulfame potassium and neotame. APM was a sweet substance occasionally found in the synthesis of gastrin-secreting hormones, after realizing that it has a sweet taste, it is then used as a sweet additive. APM is about 200 times sweeter than sucrose, and phenylalanine is produced alongside the body's metabolism. Patients with phenylketonuria cannot eat this. Therefore, products containing APM must clearly print that it contains phenylalanine on the label. An excessive intake of cyclamate can

美食城堡的安全保卫战
A War to Protect Food Safety

also cause harm to the human liver and nervous system; TGS is extracted from sucrose, and there is no evidence to prove that it is harmful. TGS is an ideal sweet additive, but it is difficult to extract and rather expensive to produce.

常见的增味剂（鲜味剂）类包括L-谷氨酸钠、5'-呈味核苷酸二钠。L-谷氨酸钠就是"味精"，而5'-呈味核苷酸二钠与味精有相同作用。这两种添加剂都可在各类食品中按生产所需适量使用，没有规定最大允许添加量，所以有人说辣条中味精味过重是有道理的。

The most common type of favoring and flavor enhancer include monosodium L-glutamate and disodium 5'-ribonucleotide. monosodium L-glutamate is MSG, and disodium 5'-ribonucleotide have similar functions as MSG. These two types of flavorings can be added to all types of food without having a limit to how much it can be added; it is quite accurate when some people say that spicy rice-sticks contain too much MSG.

防腐剂类通常包括水溶性复配糕点防腐剂（脱氢乙酸钠、柠檬酸

Food preservatives usually include water-soluble cake preservatives (sodium dehydroacetate, sodium citrate,

美食城堡的安全保卫战
A War to Protect Food Safety

钠、山梨酸），脂溶性复配糕点防腐剂（单硬脂酸甘油酯、蔗糖脂肪酸酯），以及特丁基对苯二酚。其中柠檬酸钠是最为常见的添加剂，一般由柠檬酸和小苏打或纯碱制取，其天然存在于动植物体内，所以可以认为柠檬酸钠是无毒的，可以适量使用；山梨酸是一种不饱和脂肪酸，长期大量摄入会危害肾、肝脏的健康。

虽说离开剂量谈添加剂的毒性不够客观，但在解读其营养成分表后，小微博士的脸色愈发沉重了起来。人们往往会吐槽辣条过油、过甜、味精味过重，却没想到食用一整包辣条意味着钠摄入量将超其日常推荐值的近1.4倍！食物中的钠

sorbic acid), (fat-soluble (glyceryl monostearate, sucrose fatty acid ester), and TBHQ(tertiary butylhydro quinone). Sodium citrate is the most common type of additive, usually made up of citric acid and baking soda. Since it is naturally found in animals and plants, sodium citrate is not poisonous and can be consumed in moderation. Sorbic acid is an unsaturated fatty acid, and long-term intake of large quantities will endanger the health of the kidneys and the liver.

Although speaking of the toxicity of additives is rather subjective without taking the dosage into consideration, after reading through the table of ingredient and nutrition, Dr. Micro's face darkened. People often complain about how spicy rice-sticks are too

美食城堡的安全保卫战
A War to Protect Food Safety

基本以氯化钠的形式存在，俗称食盐，而食盐摄入过多极易导致高血压，同时会影响人体对钙的吸收，造成骨质疏松，这对处于发育阶段的中小学生极为不利。

"五毛零食"因利用多种添加剂使其口味新奇，再加之这些零食的价格均在5毛钱左右，家长平日给的零花钱足以买一大堆，可谓"物美价廉"，对绝大部分中小学生极具诱惑力。但对正处于胖胖熊这一年龄阶段的孩子来说，他们更需要

sweet, too oily and too much added MSG, they actually don't realize that an entire pack exceeds the recommended sodium intake by 1.4 times! Sodium mainly exists in food in the form of sodium chloride, also commonly known as table salt. An excessive intake of salt can lead to high blood pressure, and it can affect the body's ability to absorb calcium, eventually causing osteoporosis. This is extremely bad for primary and middle school students in their stage of puberty.

These cheap snacks that costs less than a dollar are usually made with a lot of additives, which flavors the snacks with different tastes. These snacks are normally priced around 50 cents a bag, so kids are able to buy plenty with their allowance. The price is a good

A War to Protect Food Safety

营养均衡、健康卫生的食品。然而生产、销售都以散、乱、隐蔽为特点的五毛零食，不但无法保证其安全性、卫生合格与否，而且长期高糖、高盐、高热量的食用，终将导致便秘、发育不良等一系列慢性身体疾病。

看着正步步走来充满朝气的胖胖熊，小微博士坚定了内心的想法——呼吁监管部门联合校方与家长，共同抵制"五毛零食"刻不容缓。

deal to so many students loving these snacks. However, having a healthy diet while consuming safe and nutritious foods are vital for kids as the same age as Little Bear. Snacks such as these 50 cents rice-sticks can be unsafe, the expiration date alongside the production site's sanitation cannot be guaranteed. Consuming an excessive amount of sugar, salt, and fatty foods can also cause constipation and other chronic illnesses.

As Dr. Micro saw Little Bear walking towards her with such excitement, she decided that she must call up a boycott on these unsafe and unhealthy 50 cents snacks. The school, parents and authorities should cooperate together to ensure food safety for children.

美食城堡的安全保卫战
A War to Protect Food Safety

图 1 五毛零食代表——辣条

Figure 1　Representative of cheap snacks——spicy gluten sticks

【小微博士有话说】

或许会有人问，都是甜味剂，为何要同时添加那么多种？首先，甜味剂复合产生的效果绝不只是1+1=2，因此，同时添加多种可以降低成本；其次，有些甜味剂如安赛蜜、甜蜜素，高浓度时会有苦味，多种甜味剂复合后可以改善口感，同时提高甜味的稳定性；最后，若使用单一甜味剂来达到所需的甜度，其添加量很有可能超过国家标

[Dr. Micro's notes and tips]

Many people ask that since all sweeteners have the same function, what is the purpose of adding so many different kinds at once? First of all, the result of adding different sweeteners is not as simple as 1+1=2. Therefore, adding different types of additives can reduce costs. On the other hand, some sweeteners, such as Acesulfame-k and cyclamate, actually taste bitter in high

A War to Protect Food Safety

准,选择多种添加剂则可以避免超标。

不同种类的防腐剂也各有作用。水溶性的防腐剂是针对面筋的,脂溶性的防腐剂则是针对食用油的。同理,防腐剂的复配也是为了增强效果,避免单一防腐剂添加超标。

"五毛零食"虽"物美价廉",但已被多次曝光,其中食品添加剂超标,就连"网红"辣条也难逃黑

concentrations. Therefore, to improve the sweet taste and its consistency, different sweeteners are added at various levels. Since it is likely to accede the national standard for sweeteners added, using a variety of different sweeteners can also avoid a single kind exceeding the national regulation.

Different types of preservatives also have different functions. Water soluble preservatives are gluten resistant, while fat soluble preservatives are directed at cooking oils. Similarly, preservatives are also used to enhance the effectiveness of different compounds to avoid a single type of preservative exceeding the national standard's allowance.

Although these 50 cents snacks are usually a good deal and delicious at the same time, they have been repeatedly

名单。小微博士建议最好摄入天然食物，尤其是正处于发育阶段的中小学生，相比于五毛零食，瓜果蔬菜才是保卫人体健康的最强王者。

exposed for containing an excessive amount of food additives, the renowned spicy rice-sticks cannot escape from being exposed either. Dr. Micro recommends consuming natural foods, especially for students in primary or secondary schools who are going through puberty. Compared to those cheap snacks, fruits and vegetables are the best at protecting and enhancing our health.

美食城堡的安全保卫战
A War to Protect Food Safety

"毒馅"小作坊
The Mill with "Poisonous Fillings"

柔和的阳光透过窗户照射在雨燕的床上,她揉揉眼睛,扑腾着翅膀正打算出门晨练,突然看见一片黑压压的东西从窗外快速掠过。

"喂,等等……",雨燕探出窗户喊道。只见一大群的苍蝇急踩脚刹,"早啊,雨燕!我们有点事要去做,再见啦!"话音未落,他们便又匆匆地飞远了。苍蝇一族在城堡里是出了名的游手好闲,平日里只管睡觉,然而今天这么早起还要

The morning light shone brightly onto Emily the Apodidae Bird's bed through the glass windows. She opened her eyes and stretched out her wings. When Emily was fully awoken and ready for some exercise. Suddenly, she saw something that looked like a bundle of black cloud, and it quickly flew past her windows.

"Hey, wait a minute!" yelled Emily as she looked out the window. It was a huge group of flies! The flies stopped immediately after they heard Emily's voice, "Good morning, Emily! Sorry, but we are in a hurry today, no time to chat!" The flies quickly flew away.

美食城堡的安全保卫战
A War to Protect Food Safety

去办事儿，这不禁引起了雨燕的好奇心，她决定偷偷地跟着他们一探究竟。

飞了好一阵子，那群苍蝇进到了一个破旧的平房里，同时又有好几群苍蝇从里面拖出一袋袋黑色的袋子，并甩上车，动作十分熟练。"袋子里都装着些什么呢？"雨燕想上前一步解开疑惑。她飞到那房子的屋顶上，轻轻地揭开一片瓦，观察着里面的一切。只见那群苍蝇正指着砧板上那一堆即将被打包好的速冻肉馅和一只大腹便便的中年公鼠议价呢。而在房子的另一边，几个"穿戴整齐"的青年黑鼠帮手从墙根操起一把大斧头，口中喊着号

Flies in Castle City are notorious for being lazy, they usually spend their day in doors sleeping. It was such a surprise that they had something to do today! Emily was curious, and she wanted to know what the flies were up to, so she decided to secretly follow them to find out.

Emily followed the group of flies for a while, finally they arrived at a rundown bungalow. As the flies flew in, some other flies flew out of the house, carrying a large black bag, and it was later thrown onto a car. "I wonder what is inside the bag?" thought Emily, she continued to search for an answer. Emily flew on top of the roof, carefully picked out a piece of tile, and peeked into the hole to observe the room. She saw that the group of flies were bargaining with a paunchy middle-aged rat pointing at

美食城堡的安全保卫战
A War to Protect Food Safety

子,将一块块不知比他们体积大多少倍的肉一点点地剁碎。有时,肉会飞出掉到地上,旁边的母鼠便弯腰捡起,直接扔回砧板上。

由于室内卫生环境极差,源源不断地散发出令人作呕的臭味,雨燕不得不暂时离开,去呼吸一下新鲜空气缓缓。她在周边转悠了几圈,发现小作坊的附近根本看不到一个养殖场。那么这些肉都是从哪来的,又将售往哪里呢?

雨燕以一副吃货的形象和作坊附近的居民们闲聊起来。从美食城堡的美食,聊到这里的小吃,气氛

a pile of frozen meat. On the other corner, Emily saw a couple of horribly dressed black mice chopping up a huge piece of meat with an axe. Bits of meat would fly out onto the floor while being chopped, and the female mice helpers standing on the sides would pick them up and place them directly back on the chopping board.

The indoor environment was absolutely disgusting and unsanitary, the room was filled with a pungent stink. Emily even had to leave to catch her breath with some fresh air. She wandered around further and realized that there wasn't a single farm around this "mill", so where did the meat come from? Where were they going to sell the meat?

Emily started to casually chat around with the neighbors living around the mill to find out more. A lot of

美食城堡的安全保卫战
A War to Protect Food Safety

那叫一个欢快。但当谈及那个肉馅作坊时，大家顿时满脸的避讳。兔奶奶凑在雨燕耳边，轻声告诉她，"这家作坊里的老板经常会骑着辆黑乎乎的三轮车到周边收购些不干净的肉，这些肉不是我们平时毛掉不吃的"血脖肉"，就是些得病的有问题的肉。不过，我们只见他拿去加工，但从来不在附近兜售。我们也不知道它是否具有生产许可证，也不清楚加工好的肉馅会被拿去干什么。"

听到这，雨燕着急了。先不说那大打折扣的营养，"血脖肉"最大的危害来自于其上的淋巴结。甲状

neighbors joined in the conversion about delicious foods in Castle City, and finally Emily directed the conversation to the meat filling mill. Everyone suddenly went silent and tried to avoid the topic. Grandma Rabbit went up to Emily and whispered in her ear, telling her that the meat in the dirty mill was "meat waste." The owner of the mill frequently rode a tricycle and bought these unclean meat from somewhere nearby, and the meat scraps were thrown out by villagers. She saw the scrap meat being processed in the mill, but no one knew whether it's legal or not, or where the meat is being sold. One thing for sure is that none of the neighbors around will buy it.

Emily became rather anxious after she listened to Grandma Rabbit's words. Dirty scrap meat is extremely

· 15 ·

美食城堡的安全保卫战
A War to Protect Food Safety

腺。如果这些肉馅被不法商家拿到市场上进行交易，那就出大事了。她赶紧告别居民们，并打通了小微博士的电话。不一会儿，小微博士和几个警察便火速赶到，恰好堵住了那群"提货"成功正要离开的苍蝇。

经调查，这些肉馅确实是旧收购来的问题肉类加工而成。这家小作坊实则是中年公鼠的私人住所，他们并没有生产许可证，而且生产出来的肉馅也没有经过任何质量检查，就直接兜售给那些苍蝇们。再由那些苍蝇们运往美食城堡，制成肉包出售。因为这种肉包成本较低，苍蝇们能从中获取暴利，因此这家"毒馅"小作坊才"备受青睐"。

dangerous because it contains the lymph node and thyroid gland area. If the meat scrap is sold as filling on the market, then there is a big problem. She quickly warned the neighbors, and called Dr. Micro right after. In a few moments, Dr. Micro and police officers rushed to the scene. They arrived just in time to block the delivery of the dirty meat.

After an investigation, it became an evident that the meat in the mill was collected meat scrap being processed to meat fillings. The small mill is actually the middle-aged rat's private "home", and he does not have any sort of production permits. The processed meat did not go through quality check either, it was directly given to the flies to deliver. The flies carried the meat into Castle City, and

美食城堡的安全保卫战
A War to Protect Food Safety

made it into meat buns. This type of meat bun had a very low cost. Therefore, the flies could make a huge profit from selling them. This was why this small "poisonous mill" selling meat scraps was so "popular".

图 2 黑心小作坊

Figure 2 Small workshop

【小微博士有话说】

1. 血脖肉，位于动物颈脖处，含有大量淋巴结、甲状腺等。

[Dr. Micro's notes and tips]

1. Dirty scrap meat is extremely dangerous because it contains the lymph node and thyroid gland area.

美食城堡的安全保卫战
A War to Protect Food Safety

2. 淋巴结可过滤、杀灭、吞噬病原微生物和病毒等，但同时积存了大量的病菌和病毒，短时间加热不易将其杀灭，所以食用后很容易感染疾病。

3. 甲状腺的主要成分是甲状腺激素，其性质稳定、耐热，要加热到600℃以上才会被破坏。人过量食用后，大量的甲状腺激素将扰乱人体正常的内分泌活动，影响神经系统，严重者会出现各种中毒症状，甚至可能导致死亡。

4. 血脖肉并非完全不能吃，但很难将淋巴结这些有害组织切除干净，而且卖家也不愿把时间浪费在

2. Lymph nodes can filter, kill, and ingest pathogenic microorganisms and viruses, but at the same time it will accumulate a large number of bacteria and viruses that cannot be killed if it is heated for only a short amount of time, so it can cause infections if consumed.

3. The main component of the thyroid gland is the thyroid hormone, it is stable and heat resistant in nature. It can only be damaged when it's heated up to 600℃ or above. If humans consume it excessively, then the thyroid hormones will disrupt one's normal endocrine activity and disrupt the nervous system. In more severe cases, one may show symptoms of poisoning, and it may even lead to death.

4. The dirty scrap meat isn't completely inedible, it can be consumed if the lymph nodes and other harmful

美食城堡的安全保卫战
A War to Protect Food Safety

这个环节。但面对蒸熟后的"毒馅"肉包，我们又该如何接招呢？

首先，搅。血脖肉肉色苍白，即使加入大量调料也会有少许的腥臭味。如果有可能，用筷子搅拌生馅，会发现馅料过于黏稠。

其次，尝。"毒馅"包子口感黏度超出正常口感，通常也会表现得过咸、过油、过香。

最后，择。拒绝黑作坊，选择在有"食品卫生许可证"的店铺购买，这样的包子质量比较有保障。

components are cleaned out. But businesses do not want to waste time on this matter. So how can we prevent ourselves from buying the "poisonous" meat buns being sold on the market?

First of all, you can try stirring it. The dirty scrap meat will usually show a pale white color, and even though a lot of additives are added, a stench will still remain. If you try stirring the meat with a chopstick, you will also find that the meat can be overly sticky.

Secondly, through tasting. The "poisonous" fillings can have a very sticky texture, sometimes even overly salty, oily and savory.

Lastly, by choosing the right store. Please choose to purchase foods from licensed food stores, the quality of the food sold will be guaranteed.

美食城堡的安全保卫战
A War to Protect Food Safety

爱"洗澡"的草莓
The Strawberry Who Loves to Bathe

这一天美食城堡小学组织同学们课外联谊，去郊区呦呦草莓园摘草莓，让小朋友们体验采摘的乐趣。

胖胖熊、松仔、雨燕三个好朋友来到了呦呦草莓园的一个种植篷内，一个个红彤彤的草莓结在一排排草莓植株上面，看得大家直流口水。胖胖熊一看这番景象还没等带队的老师说话就直接溜进去摘了颗大草莓吞进了肚子里面。松仔看见了，立马将胖胖熊喊回来聆听带队老师讲话。带队老师看到大伙都齐了，开始说道："我们进去之后不要乱踩乱踏，要珍惜草莓园主

The day was Castle Elementary School's field trip day! All the students were going to a strawberry farm outside of the city for a fun time of strawberry picking, what a lovely day!

Little Bear, Squirrel and Emily arrived at the farm and walked into a strawberry canopy. They saw rows and rows of delicious red strawberries dangling from the strawberry bushes, what a scene! Little Bear was very excited, without waiting for the teacher's instructions, he quickly picked a big strawberry and swallowed it whole into his stomach. Squirrel saw Little Bear and dragged him back in line

美食城堡的安全保卫战
A War to Protect Food Safety

人的劳动成果，另外摘下来的草莓要洗干净后才能吃。"老师讲完话以后，同学们就都拿着小篮子进到篷内摘草莓。三个小伙伴呈一列进入篷内，胖胖熊排在第一个，将一个个又大又红的草莓放入自己的篮子，不一会儿篮子就满了。看着眼前的草莓，胖胖熊心想："这些草莓挂在草莓株上，没有掉地上，肯定都很干净，而且我刚才吃了一个并没有不好的感觉，老师一定是小题大做，我偷吃就好啦！"于是，胖胖熊趁着松仔和雨燕不注意，一边吃一边摘。松仔和雨燕也是第一次见到这么多草莓，注意力一直在怎样采到又大又好的草莓上，对胖胖熊没怎么注意。

to listen to the teacher's instructions. The teacher said, "After we enter, please be careful and try not to step on the plants, we must respect the farmer's hard work. Also, remember to wash the strawberries before eating them." After the instructions were given, all students walked into the canopy with little baskets for strawberry picking. The three friends walked in together, Little Bear walked in front of everyone, and picked many big red strawberries, his basket was quickly filled. Little Bear thought to himself "these strawberries were directly picked from the plants, they weren't on the floor, so they must be clean! I just eat one and I don't feel sick at all, the teacher is just being too careful. I'm going to eat some more!" So, Little Bear kept eating the

美食城堡的安全保卫战
A War to Protect Food Safety

大概一个小时以后,同学们陆续带着自己摘的草莓出来,洗过后就在休息区开始吃自己采的草莓。胖胖熊已经在大棚内吃了很多,不过他还是将自己带出来的草莓洗净之后和松仔、雨燕他们一起吃了起来。可是没一会儿,胖胖熊肚子就开始痛了,松仔和雨燕见到这情况就向带队老师汇报。带队老师将胖胖熊送到医院。医生询问并检查后说:"胖胖熊这种情况是由于食用大量未清洗的草莓所引起的,具体情况你们可以去问一下小微博士。"

松仔和雨燕就迅速找到了小

unwashed strawberries when the others were not paying attention. Squirrel and Emily were too focused on picking their own strawberries, so they did not pay attention to what Little Bear was doing.

An hour later, the students started to come out with their baskets, they washed their strawberries in the rest area and began eating. Little Bear had already eaten plenty, but he still washed a few and ate his strawberries with Squirrel and Emily. After a while, Little Bear's stomach began to hurt. His friends quickly reported the issue to their teacher. The teacher brought Little Bear to the hospital, after check-up, the doctor said, "Little Bear ate a large number of unwashed strawberries, you can ask Dr. Micro for the specifics."

Squirrel and Emily quickly found

美食城堡的安全保卫战
A War to Protect Food Safety

微博士。博士得知情况后就跟他们讲:"草莓是低矮的草茎植物,虽然是在地膜中培育生长,在生长过程中还是容易受到泥土和细菌的污染,所以草莓入口前一定要把好清洗关。因此,吃草莓之前一定要清洗干净或放在盐水中浸泡5分钟,以防不洁食用引起腹泻。"松仔深深地点点头,表示赞同,他们回到医院,把小微博士的话告诉了胖胖熊,说:"经过这次教训,你以后要认真听老师的话。"胖胖熊说:"好的,我一定管住自己的嘴,吃之前洗一下草莓,哈哈!"

Dr. Micro. After listening to their explanation, she said, "strawberries are plants with low stems, even though they are grown on plastic sheeting in a canopy, they're still prone to being contaminated by soil and germs. They must be washed thoroughly before eating. You can soak them in salted water for 5 minutes, or else they may cause irritation to the stomach." Squirrel nodded and agreed with Dr. Micro. After returning to the hospital, the friends told Little Bear what Dr. Micro had just explained. "See!" they said to Little Bear, "that teaches you to listen and pay attention to the teacher, Little Bear!". "Okay, I will try to control myself and wash the strawberries next time!" Little Bear surely learned his lesson.

美食城堡的安全保卫战
A War to Protect Food Safety

图 3　要洗澡的草莓

Figure 3　The strawberry who needs to bathe

【小微博士有话说】

不论任何水果，在吃之前一定要洗干净，以免发生腹泻。

[Dr. Micro's notes and tips]

No matter what type of fruit it is, please remember to wash it before eating to prevent the possibility of food poisoning and diarrhea.

美食城堡的安全保卫战
A War to Protect Food Safety

被密封的水果宝宝
The "Contained" Fruit Babies

这天城堡中心超市大打折，放学后胖胖熊拉着大家就奔向了中心超市，松仔买了几个桃子，胖胖熊则拎了三盒黄桃罐头。回家的路上，胖胖熊嫌弃松仔买的新鲜的桃子容易腐烂，放不了几天。

松仔喃喃自语道："为什么罐头里的黄桃可以保持很久不腐烂呢？难道是加了一些不让罐头腐烂的化学物质吗？"

"我们一起去图书馆查一查

Castle city supermarket had a huge sale going on, so Little Bear brought everyone to the supermarket immediately after school. Squirrel bought a couple of peaches, and Little Bear bought 3 cans of peaches. On the way home, Little Bear told Squirrel that buying fresh peaches was a bad idea, because they can easily go bad, he only had a few days to eat them.

Squirrel asked Little Bear, "Why is it that canned peaches can stay 'fresh' much longer? Is it because some sort of chemical is added to prevent it from going bad?"

"Let's go to search it up in the

美食城堡的安全保卫战
A War to Protect Food Safety

吧，刚好我们可以在图书馆把今天的作业完成后再一起回家。"雨燕开心地回答。于是，三个人开开心心地一起来到了城堡知行书店。

半个小时后，松仔和雨燕已经翻了好几本书了，胖胖熊的一盒罐头也见底了。雨燕说，"书上说罐头是真空包装，没有氧气，细菌不能繁殖，所以不会腐烂，我们自然课上老师也讲过。"

"那我懂了为什么罐头如果有胀盖就不能吃了，因为胀盖就进了空气，会有细菌繁殖。"松仔惊喜道。

library, we can also finish our homework there and go home together." Emily happily replied. So, the three friends skipped happily to the Castle City library together.

Half an hour later, Squirrel and Emily had already gone through a couple of books, and Little Bear finished a whole can of peaches. Emily said, "The books say that the cans are vacuum-packed, so there isn't any oxygen inside, bacteria cannot grow in such conditions and the peaches do not rot, and my science teacher also said something similar in class before."

"Now I know why you cannot eat canned food when the can is bloated. When there is air in the can, then bacteria will multiply!" Squirrel said with surprise.

美食城堡的安全保卫战
A War to Protect Food Safety

松仔又纳闷罐头瓶子上的添加剂"D-异抗坏血酸钠"是什么。

雨燕又找了一本关于食品添加剂安全问题的书为大家解答，D-异抗坏血酸钠是一种维生素C的立体异构体，是食品行业中重要的抗氧保鲜剂，可保持食品的色泽、自然风味、延长保质期，且无任何毒副作用。

胖胖熊兴奋地说道："看，无任何毒副作用，罐头一点害处都没有。"说着胖胖熊又打开了一盒罐头请大家吃。

松仔拿起书本说道："看这里，书上说D-异抗坏血酸钠干燥状态下，在空气中相当稳定，而在溶液中暴露于大气时则会迅速变质，也就是说我们打开罐头后要尽快吃掉的。"

雨燕看到了小微博士也在书

Squirrel then wondered what the additive Sodium D-ascorbate is?

Emily found a book on food additives to answer the question. Sodium D-ascorbate is a Vitamin C stereoisomer, an important antioxidant in the food industry, and it can maintain the food's color, natural taste, extend its shelf life, all without any side effects.

Little Bear happily said, "See! Canned goods are not harmful at all!" He opened another can to share with his friends.

Squirrel held up the book and said: "Look here, it says Sodium D-ascorbate is quite stable in dry air, but if it's in a solution exposed to air it will rapidly deteriorate, therefore we must finish the can as fast as possible."

Emily saw that Dr. Micro was at

美食城堡的安全保卫战
A War to Protect Food Safety

店，一群人便跑去打招呼。

"小微博士您也在这儿呀？"雨燕问道。

小微博士看到大家都在图书馆，便夸奖道："是啊，你们也都在啊，这是一个好习惯喔！"

"我买了罐头，他们很奇怪为什么罐头保存的时间那么长，总认为这个罐头添加了不干净的东西，所以过来查查资料，其实一点问题都没有的。"胖胖熊回答道。

"那你们都有什么收获？"小微博士问道。

松仔详细地说了他们的发现

the library too, so they all went over to say hello.

"It's great to see you, Dr. Micro!" Emily said.

Dr. Micro was happy to see that everyone was at the library, "This is a great habit, to come to the library often!"

"I bought canned goods, but they are all wondering why the food inside can be preserved for such a long time, they thought that there were harmful additives in it, so we came here to do some research. But we just found out that there was nothing harmful at all!" Little Bear said to Dr. Micro with delight.

"So, what did you all learn from your research?" Dr. Micro asked everyone.

Dr. Micro applauded everyone's

美食城堡的安全保卫战
A War to Protect Food Safety

后，小微博士连连称赞大家的探索精神并向大家解释合格的罐头里确实没有防腐剂。

胖胖熊立马接话道："看，罐头没有任何坏处，还有营养，保存时间又长，多好。"

小微博士站起来说："尽管罐头里没有添加防腐剂，但是有些不合格厂家会私下添加防腐剂，所以买罐头要去正规厂家，看合格证。另外，你们吃水果是为了什么呢？"

松仔思索了一下说道："为了补充我们体内需要的维生素。"

小微博士说道："松仔说的对，但是罐头在制作过程中却破坏

eagerness to learn after Squirrel had explained their findings, then she explained to everyone that a standard can does not contain any preservatives.

Little Bear said again, "Look, canned goods are not harmful at all! They are nutritious and they are good for a long time."

Dr. Micro continued to explain, "Although there are no additives in the cans, there are still off-grade factories who secretly adds preservatives, therefore remember to buy them from standardized factories, and check for their certificate. Also, why do you eat fruits?"

Squirrel thought for a moment and said, "to supply our body with the necessary vitamins and nutrients."

Dr. Micro replied, "Squirrel is absolutely correct, however, many of the

美食城堡的安全保卫战
A War to Protect Food Safety

了水果中富含的维生素和氨基酸，这大大地降低了其营养价值。"

说着小微博士就拿起了桌上的罐头指着添加剂甜蜜素说到："我给你们半个小时的时间探究探究这甜蜜素到底是什么？"半个小时后雨燕首先说了自己的发现：甜蜜素是一种常用甜味剂，其甜度是蔗糖的30～40倍。

松仔点点头说道："如果摄入过量的甜蜜素会对人体的肝脏和神经系统造成危害，尤其对老人、孕妇、小孩危害更明显。"

胖胖熊惊讶地说道："我以后再

vitamin and amino acid are destroyed during the production of canned fruits, so this greatly reduces the fruit's nutritional value."

Dr. Micro picked up a can from the table, and pointed to the additive sodium cyclamate on the label and said, "Can you find out what exactly sodium cyclamate is under half an hour?" Half an hour later, Emily was the first to tell everyone of her findings, "Sodium cyclamate is a type of sweet additive, it is 30 to 40 times sweeter than sucrose!" said Emily.

Squirrel nodded and added to Emily's findings, "over consumption of sodium cyclamate can cause harm to human's liver and nervous system. Seriously, kids and pregnant women are especially prone to its harms."

"I will never eat canned fruits

A War to Protect Food Safety

也不要吃罐头了"。

松仔拍了拍胖胖熊的肩膀："以后少吃点就行，还是像我一样买新鲜水果好。"

此时，最后一缕阳光洒在了书桌上，这是在和大家告别呢，大家也朝小微博士挥了挥手，满载而归。

again!" Little Bear said with surprise.

Squirrel patted Little Bear on the back and said, "It's okay, just eat less from now on and buy fresh fruits like me!"

The sun started to set, and the last ray of sunlight shined on the table, it is time for everyone to say goodbye. The kids waved goodbye to Dr. Micro after a fruitful day of new discoveries.

图 4　这么多的罐头

Figure 4　So many canned food

美食城堡的安全保卫战
A War to Protect Food Safety

【小微博士有话说】

1. 罐头制作过程破坏了水果的氨基酸和维生素。

2. 罐头虽好，老人和小孩却不宜多食用。

3. 某些工厂制作过程不合法，为了防腐会添加防腐剂，另外还会添加色素等添加剂。

4. 购买罐头注意事项：

① 选用正规厂家，品牌罐头；

② 包装完好。

[Dr. Micro's notes and tips]

1. The fruit canning process destroys the fruit's natural amino acids and vitamins.

2. Even though canned foods can be delicious, it can be harmful to children and elders.

3. Certain factories do not abide to production regulations by adding preservatives, and they may also add food coloring and other types of additives.

4. Precautions when purchasing canned food:

a. Please select factory that produced standardized cans.

b. Make sure that the packing is in place and undamaged.

美食城堡的安全保卫战
A War to Protect Food Safety

奔跑吧，外卖君
Go Deliveryman, Go!

临近饭点，美食城堡里就会出现一个特别的服务群体。他们着装统一，不管天气好坏，都奔跑在大街小巷上；他们是外卖配送员，人们眼中的美食飞行家。

随着生活节奏的不断加快，外卖行业风生水起。因为外卖省时省力，手指一动一切就都解决了；再加上食品种类繁多，可以迎合不同人的口味，所以备受上班族的青睐。

When it's almost time to eat a meal, a special group of people will appear on all corners of Castle City. They all wear the same uniforms, and they run around across the city no matter the weather. They are the delivery man, the carrier of delicious meals.

The food take-out business is booming as our pace of life continues to accelerate. Food takeout is extremely convenient, you can order all kinds of food at your fingertips. Because there is a wide variety of food online, and everyone can find food to their taste on their phones, food takeout has become

A War to Protect Food Safety

今天，熊妈妈照例给胖胖熊订了各种外卖后，就匆匆地赶去上班了。"叮咚，叮咚……"，很快，外卖就如约送达。胖胖熊按捺不住一颗吃货的心，迫不及待地全部打开，"哇！糖醋里脊、酸辣土豆丝儿、芹菜肉丸汤……都是些我爱吃的呢！"看着这些，胖胖熊的肚子"咕噜噜"闹腾地更厉害了，但他一个人吃不了这一桌美味，所以胖胖熊决定和小伙伴们一起享用，可绝不能辜负了这一桌美味。

外卖盒里色泽红亮的糖醋里脊实在是诱人，最先到达的雨燕顾不得拂去羽翼上的尘土，夹起一块便往嘴里送。"嗯……"，都来不及更完整地评价，她便夹起了第二块。可是，当雨燕又一口下去，再细细

very popular amongst office workers.

One day, Mother Bear left for work after she ordered takeout for Little Bear. "Ding dong, ding dong", the delivery arrived in no time. Little Bear couldn't wait to eat the delicious food, and ran to the doorsteps to welcome the deliveryman. "Wow! Sweet and sour pork, stir-fry potato shreds, celery meatball soup…all of my favorite dishes!" Little Bear's belly began to rumble, but he couldn't finish all the food, so he decided to share this feast with his friends!

As soon as Emily arrived, she picked up a piece of pork immediately, the glowing color of the sweet and sour pork in the take-out box was just so attractive. "Yum!" said Emily, she immediately dug in for a second piece of

美食城堡的安全保卫战
A War to Protect Food Safety

回味起来时,她感觉到酸甜之中隐约还夹杂着一股塑料味儿。"哎?难道这只是偶然的味觉感知失误吗?",雨燕决定一探究竟。她起身去厨房倒了杯白开水,在一番认真地漱口之后,雨燕分别尝试了酸辣土豆丝儿、芹菜肉丸汤等其他几道菜。这一圈儿地试吃下来,她终于解开了疑惑,问题就出在外卖身上。

雨燕发现,每份外卖里几乎都透着一股塑料味儿,或轻或重。并且,在那份现在仍是温热状态的汤里味道更为明显。雨燕皱了皱眉头,正想要告诉胖胖熊。就在此时,门"吱嘎"一声被推开了,原

pork. But, after Emily took another bite, she sensed a taste of plastic between the sweet and sour flavor of the pork. "Huh? Am I losing my sense of taste?" Emily began to wonder, and she was determined to find out what the taste was. She went to the kitchen and drank a glass of water, after thoroughly rinsing out her pallet, she decided to try out the stir-fry potato shreds, celery meatball soup, and other remaining dishes. After Emily tried out all the dishes, she finally came to the conclusion that the problem was on the delivered food itself.

Emily realized that almost every dish contained a smell of plastic, some heavier than the others. The smell of plastic was the heaviest in the lukewarm bowl of soup. Emily frowned, just as she was about to tell Little Bear of this,

· 35

美食城堡的安全保卫战
A War to Protect Food Safety

来是小微博士来了。

小微博士看到餐桌被各种形状的外卖盒子满满地霸占着，顿时面容失色，"胖胖熊，这就是你电话里的绝世美食？"她诧异地问。胖胖熊一边热情地拉着小微博士坐下，一边自信地答道："味道可棒啦！我都连续吃外卖好久了呢。"

听他这么一说，小微博士突然有些明白，为什么胖胖熊会无缘由地闹肚子了。大部分外卖商家为了利益最大化，通常组的都是小店铺，甚至有些就在公厕或垃圾场旁，环境卫生很难达标；此处，所选用的食材也很难保证新鲜。而商家为了提升口感，都喜欢用"多

the door opened and Dr. Micro walked into the house.

Dr. Micro was astonished after she saw all the different shapes of take-out container all over the table. "Little Bear, is this the feast you were talking about on the phone?" asked Dr. Micro with surprise. Little Bear welcomed Dr. Micro to sit down and said with confidence, "Yes! I have been eating delivery for a while now, and they are absolutely delicious!"

Upon hearing what Little Bear have said, Dr. Micro finally understood why Little Bear constantly had an uncomfortable stomach and frequent diarrhea. Most take-out businesses want to maximize their profit, so they normally rent out a smaller store, and sometimes even located beside a public

A War to Protect Food Safety

油、多调料"的绝招，来掩盖一些食材本身的味道。因此，长期食用这种外卖极易引起身体的不适。

"是的。虽然味道确实还不错，但大多外卖都明显偏油腻，而且还有一股塑料味"，雨燕十分赞同小微博士的话。

"塑料味应该是外卖餐盒导致的"，小微博士解释道，"目前市面上的塑料餐盒都用分级来规定耐热度"。

合格的塑料制品底部会有一

bathroom or a garbage station. The sanitary quality of the stores is hard to guarantee. On the other hand, it's hard to ensure the food's freshness and safety as well. Businesses usually use flavoring and food additives to cover up any smells that the food might have, so eating delivery food long-term may cause health problems.

"That is true. Even though the dishes may taste good, most delivered food is very greasy, it even has a plastic taste to it." Emily completely agreed with Dr. Micro's.

"The taste of plastic is probably caused by the delivery containers", said Dr. Micro, "current plastic silverware on the market has a strict heat-resistant standard."

Qualified plastic container

美食城堡的安全保卫战
A War to Protect Food Safety

个三角形符号，表示"可回收再利用"；三角形内的阿拉伯数字为"1—7"，分别代表着不同的塑料材质，主要是方便了解其使用条件。其中，标有"5"的重复利用标识，说明它的主要制作材料为PP，这种塑料容器透气性好，耐热温度高，抗多种有机溶剂和酸碱腐蚀，而且机械性质强韧，多用于微波炉餐盒、奶瓶等容器。而标有"6"的重复利用标识，表明它的主制材料为PS，常见于碗装泡面盒、快餐盒等，这种材料容器不耐高热，也不能用来盛放过酸或过碱的物质，因为它会分解出有害的聚苯乙烯。所以最好选用PP材料的餐盒，这种外卖也是唯一可放进微波炉加热食用的。

products will have a triangle symbol printed at the bottom of the product, meaning "reusable and recyclable". There is a numeric number 1 to 7 within the triangle, it represents the product's condition of use. Amongst them, the number "5" represents reusable, and that the main material used for production is PP. This type of plastic is heat resistant and air permeable, and it is also resistant to many types of organic solvent and acid based corrosion. Microwavable containers and milk bottles are usually made by such durable material. The label number "6" also represents reusable material, but its main component is PS plastic, ramen and takeout containers are often made by this plastic material. This type of material isn't heat resistant, and it

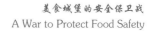

cannot contain overly acidic or alkali substances because it can breakdown into polystyrene that is harmful. Therefore, it is best to choose takeout containers made out of PP material, because it is microwavable and heat resistant.

However, the price of this type of container is usually rather high, and businesses would usually choose Polyvinyl Chloride Polyester (PVC), Polycarbonate (PC), or foam boxes instead to reduce costs. They may even choose containers made of worse materials to pack the food.

但这些餐盒的价格通常不便宜，外卖商家为了减低成本，普遍使用聚氯乙烯（PCV）、聚酯及聚碳酸酯（PC）、泡沫塑料盒，甚至是再生塑料等劣质餐盒来打包食物。

Little Bear and Emily checked every single delivery container and realized that not a single one passed the quality check. At last, Little Bear dropped his head with disappointment, "I will never eat delivery meals again."

最后，胖胖熊一脸失望地感慨："再也不敢吃外卖了"。因为他和雨燕翻看了餐桌上的所有外卖盒子，竟发现没有一个餐盒是合格的。

图 5　吃完的外卖

Figure 5　The finished take-out food

【小微博士有话说】

纯净的PP材质餐盒是十分安全的，但在实际生产过程中，很有可能被掺染了其他有害物质，仅凭标识还不能有效地鉴别。因此，小微博士支了几招鉴别"黑餐盒"的方法。

[Dr. Micro's notes and tips]

Food container made with pure PP material is very safe, but it could be mixed with harmful substances during the actual process of production. Therefore, it is hard to identify its safety by looking at the printed label itself. Dr. Micro has a couple of tips and tricks

美食城堡的安全保卫战
A War to Protect Food Safety

to identify those "containers that have gone bad".

1. 看。首先看餐盒上是否有"QS"标志及编号，再看餐盒的表面是否光洁无杂质。一般来说，颜色越深越鲜艳的餐盒越不安全。

1. Observe. Look for a QS sign and serial number on the container, then check to see if the surface of the container is smooth and without impurities. Generally speaking, the darker or brighter the color of the container is, the more unsafe it is.

2. 摸。使用再生塑料或大量添加工业级碳酸钙、滑石粉生产出来的餐盒，强度一般都很差，轻轻一撕就破。

2. Touch. the quality of plastic container made by recycled plastic, large amount of industrial grade calcium carbonate and talcum powder is generally very poor. You can easily tear and break apart.

3. 闻。"黑餐盒"会有一股异味，而合格餐盒往往没有。

3. Smell. Unsafe containers usually have an unusual smell, while containers that have passed inspection do not.

此外，小微博士友情提醒：外卖虽然便利，但长期食用容易导致

Other reminders from Dr. Micro: Although food delivery is rather

A War to Protect Food Safety

营养不均衡，高油高脂的饮食也易引发身体多种疾病；而且，一次性的外卖盒子会造成环境的破坏与污染。因此，小微博士呼吁大家，少点外卖勤动腿，均衡饮食多运动，身体才会更健康。

convenient, it may lead to unbalanced nutrition in the long run. Besides, foods prepared in a restaurant usually have high levels of oil and fats, it may cause a variety of diseases and it is harmful to our health. On the other hand, food delivery is extremely harmful for the environment because of all the disposable containers that are being used, this will damage our environment and increase pollution levels. Therefore, Dr. Micro wants to remind everyone to go out more and order less. Be sure to maintain a balanced diet and do plenty of exercise for your body and health.

美食城堡的安全保卫战

A War to Protect Food Safety

变身芽芽的士豆君

The Potato with Tails

在美食城堡小学的一节美食课临近下课时，牛牛老师给同学们布置了一个任务，就是每个人第二天从家里面带一种食材，可以是蔬菜或水果等。放学回家的路上，松仔、胖胖熊和雨燕三个小伙伴开始讨论牛牛老师今天给大家布置的任务。松仔说："你们有没有想好要带什么？"胖胖熊和雨燕异口同声地说："没有。"松仔紧接着说："既然这样，我们一会儿回家后就看看家里面有什么多的食材，然后微信群里联系一下，咱们三个带不同的食材过去，这样可以相互使用对方的食材了。"胖胖熊和雨燕听了之后纷纷点头，表示赞同。回到家

Castle Elementary School's culinary arts class was about to end for the day, Mr. Cow gave everyone an assignment to bring a type of ingredient from home, it could be anything such as vegetable or fruit. The three friends Emily, Little Bear and Squirrel started to discuss about their assignment from Mr. Cow on their way back from school. "Did you guys know what you're going to bring to class?" asked Squirrel. Little Bear and Emily answered together "Not yet!". "Why don't we go home and see what type of ingredients we have, then let's contact with each other and talk about it. We should decide on 3 different

美食城堡的安全保卫战
A War to Protect Food Safety

后，三人查看各自家中的食物，经私下讨论后，就决定了第二天要带的食材。

到了第二天的美食课，牛牛老师走到同学们的面前说："现在请大家把自己带来的食材放在自己的桌前。"松仔带来了一袋美味的松果，胖胖熊带了几颗甜甜的玉米，雨燕带来了好几个土豆。牛牛老师一一扫视过去，最后将目光停留在了雨燕带来的几个土豆上面。仔细一看，发现这些土豆长了芽，而且土豆的有些部分还变绿了。牛牛老师立刻走到雨燕的身旁说："雨燕，现在这个土豆发芽变绿是不能食用

types of ingredient." Emily and Little Bear both nodded in agreement to Squirrel's brilliant idea. After they all went home, they checked their pantry and discussed on their options. Finally, they have decided together the ingredients they were bringing for the next day to school.

On the second day during class, Mr. Cow walked to the front of the class and asked everyone to put the ingredients they've brought to class on their desks. Squirrel brought a delicious bag of nuts, Little Bear brought out kernels of corn, and Emily had a couple of potatoes. Mr. Cow walked pass everyone's desk to check on all the ingredients, he suddenly laid his sight on Emily's potatoes. After he took a closer look, Mr. Cow realized that these potatoes all sprouted, and

美食城堡的安全保卫战
A War to Protect Food Safety

的。"雨燕吃惊地问:"为什么?"牛牛老师挠挠头说:"这个我们还是请教一下小微博士吧!"

不一会儿牛牛老师就请小微博士过来了。小微博士听完事情的经过后,就向大家解释道:"土豆的致毒成分为龙葵素,是一种弱碱性的生物碱,可溶于水,遇醋酸易分解,高热,煮透可解毒。发芽土豆或未成熟、青紫皮的土豆所含龙葵素增高数倍甚至数十倍。龙葵素具有腐蚀性、溶血性,并对运动中枢及呼吸中枢产生麻痹作用。"雨燕听到小微博士这么说,就准备把土豆扔掉。小微博士立刻阻止雨燕这

some parts of potatoes also turned green. Mr. Cow walked to Emily immediately and said "Emily, your potatoes are sprouting and it turned green, this cannot be edible." Emily is rather surprised, "why not?" she asked. "Why don't we ask Dr. Micro?" said Mr. Cow as he scratched his head lightly.

Mr. Cow went to invite Dr. Micro to class. After she had heard about the whole story, she started to explain, "Solanine is the part of the potato that is poisonous. Solanine is a type of biological glycoside that is weakly basic. It is water soluble and it can resolve in acetic acid, you can detoxify it through boiling it thoroughly in high temperature water. Potatoes that has sprouted or is unripe with green of purple peel, then the solanine concentration is ten times

美食城堡的安全保卫战
A War to Protect Food Safety

么做，并且说："我看这些土豆发芽较少，还是可以吃的。但是应彻底挖去芽的芽眼，并扩大削除芽眼周围的部分，这种土豆不宜炒吃，应该充分煮、炖透。烹调时加醋，可加速对龙葵素的破坏。"

雨燕听完博士的话后，还是有些犹豫，松仔看出雨燕的顾虑就接着博士的话说："既然博士这么说了，我相信就不会有问题。我们现在就把这些土豆的芽挖掉，然后用热水把土豆煮熟煮透，再加一些醋，相信这样会把土豆里的毒素解除。最后我们把煮熟的土豆弄成泥，搭配我和胖胖熊带来的松果和玉米一定会很好吃的。"雨燕听松

higher. Emily decided to discard these potatoes after heard the words from Dr. Micro. But she was stopped. "These potatoes remains edible from my point," said Dr. micro, "Since they are poorly sprouted, we can detoxify them by removing potato buds completely along with their surroundiny area and boiling them thoroughly with vinegar to accelerate the degradation of solanine.

After listening to Dr. Micro's explanations, Emily was still rather hesitant and Squirrel noticed her worries, "If Dr. Micro said so, then I am sure it is fine," said squirrel, "Why don't we cut off the parts with sprouts, and boil the potatoes thoroughly. We can then add vinegar to the potatoes to help detoxify it. Then we can make mashed potato out of it and add the corn and

美食城堡的安全保卫战
A War to Protect Food Safety

仔这么说，于是就按松仔说的方法做土豆，不一会儿在松仔和胖胖熊的帮助下，完成了松仁土豆泥，搭配胖胖熊的玉米粒，几个人开开心心地吃着这份美食。

nuts that Little Bear and I brought. I am sure it will be delicious!" Emily listened to Squirrel, and she began to prepare the potatoes as Squirrel has told her. Soon, with the help of Little Bear and Squirrel, they have completed the plate of mashed potatoes containing nuts and corn. The three friends finished the dish they have worked together happily.

图 6 出芽芽的土豆

Figure 6　The potato with tails

美食城堡的安全保卫战
A War to Protect Food Safety

【小微博士有话说】

1. 土豆应低温、避光贮藏，防止生芽。

2. 发芽较少的土豆应彻底挖去芽的芽眼，并扩大削除芽眼周围的部分，这种土豆不宜炒吃，应充分煮、炖透。

3. 烹调时加醋，可加速对龙葵素的破坏。

[Dr. Micro's notes and tips]

1. To prevent potatoes from sprouting, store them in low temperature and avoid sunlight.

2. Cut out the sprout and its surroundings on a potato, it is better to boil it thoroughly before eating rather than stir-fry.

3. Adding vinegar to it can help break down the Solanine.

美食城堡的安全保卫战
A War to Protect Food Safety

常吃爆米花，肥胖和你不分家
Too much Popcorn Makes you gain a lot of weight

春去秋来，花开花落。时光在不经意间悄悄流逝，美食城堡里的伙伴们就这样幸福快乐地生活着。一天又一天，一年又一年，城堡里很多事物都在不停地变化，但是也有一些东西一直没变，比如胖胖熊，还是熊如其名，拖着笨重的身体，一如既往地好吃、爱制造麻烦。可是，最近有一件对胖胖熊来说很期待的事情，那就是他的生日快到了。每年生日，熊妈妈熊爸爸都会给他做很多好吃的，还会带他去好玩的地方，最重要的是熊妈妈曾承诺每年在胖胖熊生日的那天，都可以满足它一个小小的愿望。时间过得很快，胖胖熊的生日越来越

Time passes by rather quickly when you're living happily and sound. As the days pass, many things in Castle City have changed. However, some things never changed, Little Bear was still the plump troublemaker who loved to eat. Lately, the thought of one thing was on Little Bear's mind— his upcoming birthday. Every year on his birthday, Mother Bear and Father Bear would make delicious food and take Little Bear to fun places. Most important of all, Mother Bear would grant Little Bear a wish on this day. Little Bear's birthday was just around the corner, and he had already decided what he wanted for his birthday— a day

美食城堡的安全保卫战
A War to Protect Food Safety

近了，不过他早就已经想好自己的小愿望了。原来胖胖熊的愿望是生日那天要妈妈带他去电影院看《大鱼海棠》。

胖胖熊的生日终于到了。当清晨的第一缕阳光照进城堡里，胖胖熊就睁开了眼，今天他特别开心，终于能去看电影了，而且回到家还能吃到好多好吃的东西。吃过中饭，胖胖熊和熊妈妈前往电影院。一路上胖胖熊暗暗高兴，一会儿要看电影，那就一定有爆米花吃了，爆米花可是他最爱吃的食物之一。可是胖胖熊担心妈妈忘记，于是提醒妈妈，"妈妈，你说看电影的时候一定不能少了什么呀？"熊妈妈怎会不了解这个爱吃鬼的小心思呢，于是妈妈提醒胖胖熊，他已经够胖了，是不是该少吃一些油炸食物呢？可是妈妈还是拗不过胖胖

at the cinema with Mother Bear, watching Big fish & Begonia.

Little Bear's birthday finally came. Little Bear woke up when the first ray of sunlight shone through his windows. He was finally able to go to the movies, and could feast on delicious food when he came back home. After lunch, Little Bear and Mother Bear went to the movie theater. Little Bear secretly rejoiced at the idea of a large box of popcorn— one of his favorite snacks. He was afraid that Mother Bear would forget about buying popcorn entirely, so he started dropping hints, "Mother, what is the one important thing that you cannot miss out at a movie?" Of course, Mother Bear knew Little Bear was talking about

美食城堡的安全保卫战
A War to Protect Food Safety

熊的哀求,不得不答应给他买。电影院附近卖爆米花的商店真不少,而且爆米花还有很多种口味的,有巧克力爆米花、奶油爆米花、芥末爆米花、草莓爆米花、香料盐爆米花……胖胖熊挑了一大桶奶油爆米花,这是他最爱吃的口味,然后跟着妈妈走进了电影院。

电影开始了,胖胖熊更多的心思不是在电影上,他呀,一心在吃爆米花。今天真巧,松仔也来看电影,还就坐在胖胖熊不远处。他一眼就看到了正抱着一大桶爆米花吃的胖胖熊。松仔叹了口气摇摇头,想着:"这胖胖熊难怪小小年纪就这么胖,原来是喜欢吃爆米花,今

popcorn, but she reminded him that he was already overweight and should eat less deep-fried food. However, Little Bear begged her for popcorns, and she eventually gave in. There were several stores and many flavors of popcorn to pick from. There were plenty of flavors: chocolate, butter, mustard, strawberry and salted spices. Little Bear bought a large box of buttered popcorn, which was his favorite, and followed Mother Bear into the movies.

The movie started, but Little Bear was too busy eating his popcorn to pay attention to the movie. The squirrel happened to be at the cinema that day too, and he sat close to Little Bear. Mr. Smarts sighed at the sight of Little Bear gobbling down a big box of popcorn. He thought to himself, "No wonder Little

美食城堡的安全保卫战
A War to Protect Food Safety

天被我发现了，我不能坐视不管呀"。再看胖胖熊那吃爆米花的速度，估计这电影还没完，这一大桶爆米花就要消失在他眼前了，那又得多增肥好几斤吧，不行，我得阻止他。松仔盘算着："现在在电影院，大声讲话交流是很不礼貌的，我该怎么办呢？"正想着，发现胖胖熊好像要起身出去，他连忙跟上去。原来胖胖熊是想去外面买点喝的。正当胖胖熊买好饮料，却被大尾巴松鼠叫住了。松仔告诉胖胖熊："吃太多爆米花是不好的，你不知道吃太多很容易胖的吗？"胖胖熊反驳道："可是爆米花好吃，我喜欢吃"。"你知道你为什么这么胖了吗，因为这东西脂肪含量很高的，常吃爆米花的人会肥胖，有很大原因来自于它。"可是胖胖熊却觉得松仔说得没有根据，一点都不相信

Bear is overweight, he likes popcorn! I bet Little Bear would finish that whole box before the movie ends, and that'll make him gain more weight. I should do something to help him! But it is rude to speak loudly during a movie, what should I do?" When Mr. Smarts was busy thinking to himself, Little Bear got up and headed outside. Mr. Smarts followed behind him. After Little Bear got himself a coke and was about to re-enter the movie, Mr. Smarts tapped him and said, "Do you know that too much popcorn is bad for you and it makes you gain a lot of weight?" Little Bear argued, "But I like popcorn, they're tasty!" "Now you know why you are overweight? Popcorns contain a lot of fat, after you eat all those fat, you will gain weight!" Little Bear did not believe what Mr.

A War to Protect Food Safety

他，反驳道："我知道自己胖，但这是因为我爱吃又不喜欢运动，跟爆米花没关系。"松仔觉得自己说服不了他，只能请小微博士出面了。他对胖胖熊说道："我说的是对的，你不相信，我们可以去找小微博士，我只是想帮你，并不是不许你吃。"胖胖熊答应一起去找小微博士评评理。

于是他们抛开还没看完的电影，直接奔向小微博士家。小微博士一见他俩来，就知道肯定是淘气鬼胖胖熊又有难题了。问过究竟后，小微博士心平气和地告诉胖胖熊："爆米花可以吃，但是得少吃。松仔说的都是对的，因为爆米花制作过程中使用的奶油是富含反式脂

Smarts just said. "I know I'm a little overweight, but that's only because I love to eat and I don't exercise. It has nothing to do with popcorn." Mr. Smarts knew he could not persuade Little Bear, so he decided to seek for Dr. Micro's help. "You don't believe my words, why don't we pay Dr. Micro a visit and see what she has to say? I didn't mean to stop you from eating popcorn, I just meant to help, that's all." Little Bear agreed to go to Dr. Micro's.

The two went straight for Dr. Micro's house. As soon as Dr. Micro saw the naughty Little Bear, he knew Little Bear needed some help again. After she heard what had happened, Dr. Micro turned to Little Bear and said, "You can have popcorn, but you should eat a lot less. Mr. Smarts is right. Butter in

美食城堡的安全保卫战
A War to Protect Food Safety

肪酸的氢化植物油，这种油容易使人发胖的。其实呢，爆米花还有其他危害，爆米花里含有铅。铅进入人体后，渐渐积累增多可能会损害人的神经系统和消化系统，对于儿童来说，常吃爆米花的话，容易造成慢性铅中毒，表现出食欲下降、腹泻、生长发育缓慢等现象，所以爆米花其实是很不健康又危险的食物，得少吃。"胖胖熊被吓呆了，他只知道爆米花好吃，却不知道它原来也是这么危险的东西。他告诉小微博士，自己以后会少吃的，还向松仔表示了感谢。小微博士欣慰地告诉胖胖熊："嗯嗯，你会更健康的。"

the popcorn is hydrogenated vegetable oil which contains rich trans-fatty acids. This kind of butter causes people to gain weight. Popcorn is bad for our health in other ways, too. Lead appears in popcorns, and when lead enters the human body, it accumulates and damages the nervous system and the digestion system. If children eat too much popcorn, they may experience loss of appetite, diarrhea and stunted growth, all symptoms of chronic lead poisoning. Popcorn is an unhealthy and dangerous food, and that is why I suggest you to eat less of it." Little Bear was shocked. He only knew that popcorns were delicious, but he did not know how dangerous eating popcorn could be! Little Bear told Dr. Micro that he would eat less in the future, and he thanked Mr. Smarts for his

A War to Protect Food Safety

concern. Dr. Micro was glad, "You know if you eat less popcorn, you will become healthier, Little Bear!"

图 7　香喷喷的爆米花

Figure 7　Delicious popcorn

【小微博士有话说】

1. 制作爆米花的铁罐内有一层含铅的合金，当给爆米机加热时，其中的一部分铅会变成铅蒸汽进入爆米花中，铅就会随着爆米花进入

[Dr. Micro's notes and tips]

1. The containers that makes popcorn contains a coat of lead alloy in the cans. When the popcorn machine is heated up, a portion of the lead may enter the

· 55

美食城堡的安全保卫战
A War to Protect Food Safety

人体中，长期大量食用爆米花，容易造成肺部的损伤，易引起呼吸困难和哮喘，严重的甚至危及生命，故不宜多食。

2. 爆米花或米花糖含铅量高，铅进入人体会损伤神经、消化系统和造血功能。儿童对铅的吸收比成人高数倍，加之儿童的解毒功能弱，若常食之，易致慢性铅中毒，引起食欲下降、腹泻、烦躁、牙龈发紫和生长发育迟缓等问题。

popcorn and enter our bodies alongside the popcorn when we eat it. Eating popcorn long-term may damage our lungs and cause respiratory problems, even asthma. It may be life-threatening, so please eat it less.

2. Popcorn and rice candy contains a high amount of lead, and lead that enters the human body can damage the nervous and digestive systems, even the hematopoietic function. Children absorbs lead several times higher than adults, and it can weaken children's detoxification function. If it is consumed often, then one is prone to chronic lead poisoning, diarrhea, loss of appetite, irritability, purple mouth gums and even stunted growth.

美食城堡的安全保卫战
A War to Protect Food Safety

腐臭豆腐和致癌咸菜的pk赛
Stinky Tofu VS. Pickled Vegetables

最近一段时间整个美食城堡都比较安静，没有什么特大新闻。这天，天气晴朗，万里无云。放学了，胖胖熊走在回家的路上，一边哼着老师刚教的小曲儿，一边盘算着今天该吃什么好呢？想着想着突然有一股奇怪的气味扑面而来，胖胖熊连忙捂住鼻子，嘀咕道："好臭呀，好臭呀，是什么气味呀？"他沿着臭味飘来的方向走去，听到不远处一位狐狸大叔正吆喝着："卖臭豆腐了，卖臭豆腐了，又香又美味，闻起来臭、吃起来香的臭豆腐，独家秘制哦……"。胖胖熊发现围在臭豆腐店门口买臭豆腐的小伙伴还不少，胖胖熊纳闷了：

Castle City had been quiet for the last couple of days, there were no major news stories. One day was a lovely day with a blue and clear sky. As Little Bear walked back home from school, he noticed a pungent smell. Little Bear quickly covered his muzzle and muttered, "What is this smell, it smells so bad." He followed the scent and saw uncle Fox making stinky tofu. "Stinky tofu for sale! Gourmet and it tastes great!" yelled uncle Fox. Little Bear observed that there were quite a few people surrounding the Stinky tofu stand. "It smells terrible, can you actually eat it?" wondered Little Bear.

美食城堡的安全保卫战
A War to Protect Food Safety

"这么臭的东西真的能吃吗?"

可是,看到买到的小伙伴们都吃得津津有味,胖胖熊馋得口水都快掉下来了,脚步开始不听使唤地往臭豆腐店门口挪动。这时雨燕刚好从天空飞过,臭豆腐店门前热闹的气氛和那强烈的气味引起了她的注意。她压不住好奇心,刚想停下来看个究竟,就看到了拥挤的动物群中的胖胖熊。雨燕愈发好奇了,开始向周围的伙伴询问到底发生了什么。

经过一番了解,雨燕知道了原来是最近美食城堡先后开张了两家店,卖的都是城堡以前没有或者很久没卖过的食物。其中一家就是眼前这家臭气熏天的臭豆腐店,还有就是对面一家号称是高级咸菜专卖店。雨燕觉得这突如其来的奇怪店铺不太对劲。她决定去找智慧星——松仔问问。再说这边馋嘴的

But he saw many people eating the tofu with such satisfaction, Little Bear started drooling himself and walked towards the restaurant. At the same time, Emily flew over the scene and was curious about the stinky tofu as well. She saw Little Bear squeezing pass the crowd of people trying to get the tofu stand. Emily started to ask around to find out what was happening.

After she figured out the situation, Emily found out the reason for all the hype, and it was because of two recently opened restaurants. One of which was a stinky tofu restaurant, and the other was a pickled vegetable restaurant, which were just across the street from each other. Emily found it weird that the restaurants opened so suddenly, so

美食城堡的安全保卫战
A War to Protect Food Safety

胖胖熊，因为身体比较庞大又笨重没挤得进去，所以没有买到吃的。他耷拉着脑袋一副不高兴的样子，嘴上还嚷道："不吃这陌生又臭臭的玩意儿了。"他刚转身准备回家，却又被对面吃喝的声音吸引住了。胖胖熊发现对面也新开了一家店，"咦，城堡里怎么突然多开了两家店呀，那边肯定也有好吃的"，他想到。对面店铺前的顾客也好多，这使胖胖熊更加好奇了。他好不容易才挤进动物群，映入眼帘的是各种包装得看起来像烂掉的非绿色小菜。旁边的老板却喊道："美味可口的咸菜，拌饭香香，早餐配粥，美味十足。"胖胖熊想着："妈妈和我早餐都最喜欢喝粥，我要买回家，妈妈一定会开心的。"

他正准备上前买，却被及时出现的松仔拦住了。胖胖熊不明白

she decided to find Squirrel to find out the reason. Little Bear tried to squeeze through the crowd of people to buy himself some tofu, but it was very hard as he was rather large and clumsy. "I don't want such smelly food anyway." Said Little Bear bitterly. Just when he was about to go home, his attention was caught by another voice, and Little Bear noticed the restaurant just across the street. "Wow, I didn't know that Castle Town had two newly opened restaurants, this was a surprise." Said Little Bear, "There must be tasty food over there as well."

Just when he was about to buy a package of the pickled vegetable, Squirrel

A War to Protect Food Safety

了，侧着脑袋问道："干吗拦我，我想买咸菜回家，妈妈会开心的"。松仔连忙说："你呀，就知道吃。之前因为乱吃东西吃坏肚子，你忘了，虽然近段时间来，城堡里的美食在不断变得更加健康与营养，但这突然出现的新食物，城堡以前都没有卖，我们都不了解它们的来源和成分，而且它们都那么奇怪，一个又臭又是油炸的，另一个简直就是嗖掉或者烂掉的蔬菜嘛。我得先了解了解它们，看看是否卫生和健康，我们去请教小微博士吧"。雨燕摸着胖胖熊的头说道："松仔说得对，我们不能随便吃这些来历不明的食物。"胖胖熊只好妥协，决定和大家一起去找小微博士。

小微博士得知此事后，开始了小讲堂式的讲解。她告诉大家：臭豆腐是发酵的豆制食品，发酵过

came over and stopped him. Little Bear was confused and asked, "Why are you stopping me, I want to buy a pack of pickled vegetables for my mother." Squirrel replied, "Can't you be more cautious of what you eat. The food in Castle Town is becoming healthier and now suddenly there's two new food items that smell bad or look rotted. We should know how it was produced first. Let's ask Dr. Micro." "Squirrel is right." said Emily, "We shouldn't eat food from unknown sources. Let's go to find Dr. Micro then."

After Dr. Micro heard the story, she explained, "stinky tofu is made

美食城堡的安全保卫战
A War to Protect Food Safety

程中会产生一些有害化学物质，所以会产生一股臭味，并且臭豆腐也是油炸类食物的一种，其油脂含量高，多吃对健康并无益处，小孩子最好不要吃。咸菜是由放了很多食盐的蔬菜经过很长时间浸泡而成的，如果咸菜没腌透的话，会产生亚硝酸盐。咸菜只能偶尔食用，如果长期贪食，则可能引起泌尿系统结石。另外咸菜腌制过程中，维生素C被大量破坏，而且咸菜中含有的亚硝酸盐有致癌作用，所以这两种食物都要少吃，小孩子长身体的时候最好不要吃。大家听完都震住了，大尾巴松鼠说："既然这两种食物都最好不要吃，那么我们得赶紧告诉美食城堡的小伙伴们，这里那么多孩子，大人还不要紧，小孩子不行呀。"于是大家决定即刻前往美食街，可是当他们来到美食

through the process of fermentation, so oftentimes through the fermentation process, harmful chemicals could arise to produce the smelly scent. Stinky tofu is also a fried item with a high concentration of fat, so eating a lot of stinky tofu would not be beneficial to health, especially for kids. Submerging vegetables in salt for a long period of time makes pickled vegetables, but if the vegetable is not pickled thoroughly, it will produce nitrates. Occasional consumption of pickled vegetables is fine, but long-term consumption could cause symptoms such as urinary tract stones. Also, throughout the pickling process, vitamin C in the vegtable is damaged, and might contain nitrite, which could lead to cancer. Therefore, these two food items should be eaten

美食城堡的安全保卫战
A War to Protect Food Safety

街却发现根本挤不进去呀，新开的两家店吸引了不少顾客，他们正买得火热。两家老板大声叫卖着，看那阵势是要比一下谁家销售情况更好吧。松仔见此情况，提议大家奔走相告，给城堡里的伙伴们普及一下新学的知识。胖胖熊摸摸头说："我也要参加，我要去告诉小伙伴们"。大家就这样开始想各种办法，尽量转告到城堡里每一个小伙伴。

with care, and shouldn't be given to children that are still developing. Everyone was stunned by all the information. Squirrel said, "if the two kind of food are so bad, we should spread the words and warn them about the harms of eating such food." So, they decided to go back to the restaurants. But when they arrived, the restaurant was so crowded that they couldn't even get inside. Squirrel then suggested the group to split up and spread the words by telling friends and family. "I will do that too." said Little Bear, and the friends started to spread the knowledge of these types of food to all of their friends.

【小微博士有话说】

1. 臭豆腐发酵前期是用毛霉菌种，发酵后期易受其他细菌污染，

[Dr. Micro's notes and tips]

1. Stinky tofu is fermented with

美食城堡的安全保卫战
A War to Protect Food Safety

图 8 "臭气熏天"的臭豆腐
Figure 8 "Stinky smell" of the stinky tofu

其中还有致病菌，因此过多食用会引起胃肠道疾病。臭豆腐发酵过程中会产生甲胺、腐胺、色胺等胺类物质以及硫化氢，它们具有一股特殊的臭味和很强的挥发性，多吃无益。

2. 咸菜里面有很多致癌的物质，因为咸菜的制作过程要放很多

Mucor, the fermentation process is susceptible to other bacterial contamination including pathogens. Therefore, over consumption may lead to gastrointestinal disease. Stinky tofu's fermentation process will produce methylamine, putrescine, serotonin, amines and hydrogen sulfide, these chemicals usually have a special smell that is rather volatile, and it is best not to eat it.

2. Pickled vegetables contain many

美食城堡的安全保卫战
A War to Protect Food Safety

食盐，而且要经过很长时间的浸泡，随之就会产生致癌物质，咸菜没腌透的话，还会产生亚硝酸盐，食用这种物质会导致癌症，所以为了远离癌症，一定要注意。经常吃咸菜还会导致高血压，咸菜里面食盐很多，因而钠的成分有很多，人体如果摄入过多的钠就会导致高血压，所以不要长期食用咸菜，平时吃饭也不要吃得太咸。经常吃咸菜，还会加速皮肤老化，因为盐里面的钠离子和氯离子会导致面部细胞失水，从而降低皮肤质量。

carcinogenic substances, it is because a lot of salt is added during the pickling process, and it needs to be soaked for a long time. The carcinogenic substances are produced during the pickling process. If the pickles are not completely pickled, then nitrite will be produced and consuming it will lead to cancer. Eating pickles often can also cause high blood pressure. Please do not consume pickled vegetables long-term because it is high in sodium, and an excessive intake of sodium can cause hypertension well. Eating pickles often can also accelerate the aging of skin, because the sodium and chloride ions in salts can cause facial cells dehydration, thus reducing the quality of your skin.

美食城堡的安全保卫战
A War to Protect Food Safety

翻滚的麻辣烫君

Rolling Malatang

自从班级创建了学习兴趣小组后，班级的学习氛围就更加浓厚了。这天放学后，在学校的草坪上，胖胖熊、雨燕还有松仔在专心致志地学习，他们时而阅读时而讨论，金色的夕阳恋恋不舍地收敛了最后一缕余光，像是不忍打扰到他们似的。夕阳西下，胖胖熊伸了伸懒腰，便喊着大家一起去吃晚饭。松仔邀请大家去他家里吃饭，胖胖熊摇摇头道："前几天在中心街道开了一家新的小吃店，特别好吃，我们去吃吧？"雨燕停在胖胖熊的肩膀上表示赞同。

Ever since the school started extracurricular classes, the class's atmosphere had become more studious. One day after school, Little Bear, Emily and Squirrel were lying on the lawn, reading and chatting. As the sun started to set, Little Bear did a big stretch and asked the friends if they wanted to have dinner. "How about you guys come to my house for dinner?" Asked Squirrel. Little Bear shook his head and said, "Let's go somewhere else, I heard about a new restaurant, and that have really tasty food, should we go there?" Emily agreed and so the friends went to try it out.

美食城堡的安全保卫战
A War to Protect Food Safety

三人收拾完书包就走向了中心街道，还没到店铺门口就远远地看到门口排起的长队，而且送外卖的小哥也忙得不可开交，大伙不由地加快了速度。然而到地方才发现这只是一家麻辣烫小吃店，店内烟雾缭绕，门口的锅内翻滚着的滚烫油水，散发着阵阵香味，胖胖熊流着口水便拉着大家往里面走，松仔则想起了上次吃麻辣烫时，嘴上烫了一个水泡好久都没有消下去。坐下来后松仔提醒大家待会一定要吃慢一点避免烫伤自己，雨燕和胖胖熊则满不在乎。热腾腾的麻辣烫终于端上来了，胖胖熊美美地喝了一口汤，雨燕也学着胖胖熊的做法吃了起来，突然雨燕叫道自己被烫了一下，于是她赶紧灌了自己几口凉水便接着吃，而胖胖熊也被其中的辛辣味呛得不轻。等到大家离开时，

The three packed up their bags and went to the town center to have their meal. Arriving at the restaurant, they found it packed and with a long line waiting. They walked closer and found that the restaurant was in fact a malatang restaurant. Inside smokes and mists flurried around as the pot boils outside that sent the delicious spicy aroma to the three friends. Little Bear drooled and dragged the other two inside as Squirrel remembered the boil he got in his mouth last time when he ate the same cuisine. Sitting down to eat, Squirrel reminded everyone that one should eat malatang slowly to avoid burning themselves, but Emily and Little Bear ignored the reminders. The hot and delicious malatang was finally served. Little Bear drank a spoonful of

A War to Protect Food Safety

雨燕感觉自己嘴里起了一个大水泡,而且越来越疼,三人不得不去中心卫生所对雨燕的伤口进行处理。

值班的医生看到雨燕的水泡,语重心长地说道:"你们这是吃了什么东西,烫得这么厉害,你最近的饮食可要注意了,一点都不能吃辛辣食物。"雨燕口齿不清地告诉医生他们吃了麻辣烫,医生叔叔叹了一口气说道:"你们这些孩子就喜欢吃这些东西,这些东西从养生的角度上来说是存在很多安全隐患的,你看你的嘴都烫成这个样子了,而

the broth and started eating, and Emily as well. Suddenly, Emily cried out in pain as she got burned in the mouth, she quickly drank few glasses of cold water and continued eating. Little Bear almost couldn't handle the spiciness. After the meal, as they were walking back home, Emily felt a large boil inside her mouth, and it hurt more and more. So, the three friends went to the clinic to treat Emily.

The doctor saw Emily's boil, and said, "what did you guys eat? Your mouth was burned and you should be cautious of what you eat, and for now avoid eating hot and spicy food." Emily told the doctor that she ate malatang, and the doctor sighed, "you kids all like to eat this kind of food, from a health perspective this is very unsafe. Look at your mouth, burned like this, and your

A War to Protect Food Safety

你的肠胃粘膜可更脆弱呀！"处理完之后，医生叔叔拿起书桌上的养生书给雨燕他们看，书上对于麻辣烫提到了以下五点。

1. 食材含非法添加剂：受利益驱使，不少摊贩或海产品零售商为了使海产品看上去新鲜、保存时间长，常常会使用国家禁用的工业碱、福尔马林发泡海产品。

2. 过度刺激肠胃：麻辣烫的口味以辛辣为主，虽然能很好地刺激食欲，但同时由于其过热、过辣、过于油腻，对肠胃刺激很大，过多食用有可能导致肠胃出现问题。

3. 菜煮不熟，易引起消化道疾

gastrointestinal mucosa will become weaker!" After treating Emily's mouth, the doctor took out a book, which had a few cautionary details about malatang, and showed it to Emily. The details are:

Food ingredients containing illegal additives: driven by profit, many seafood vendor and retailers use industrial soda and Faure Marin that are banned, in order to make seafood look fresh and possible for long-term preservation.

2.Excessive stomach irritation: malatang is mainly spicy, although spices are able to stimulate the appetite, it is also too greasy and overheated. Therefore, malatang can irritate the stomach and over consumption may even cause severe stomach problems.

3.Eating vegetables that are not

美食城堡的安全保卫战
A War to Protect Food Safety

病：街边麻辣烫常常是满满的一锅，如果没有烧开、烫熟，病菌和寄生虫卵就不会被彻底杀死，食用后容易引起消化道疾病。

4. 温度过高，伤害胃肠黏膜：人的口腔、食道和胃黏膜一般最高只能耐受50℃至60℃的温度，太烫的食物会损伤黏膜，导致急性食道炎和急性胃炎。

5. 口味过重：麻辣烫太浓太辣，成分过于肥腻，容易导致高脂血症、胃病、十二指肠溃疡等疾病。

医生叔叔又说道："虽然我们

thoroughly cooked can lead to digestive diseases: malatang sold on the street is usually steamed in a very large pot. If it is not completely boiled, then bacterial and parasites will not be completely killed and this can easily lead to digestive tract diseases.

4. High temperature can hurt the gastrointestinal mucosa: Our mouth, esophagus and stomach mucosa, generally speaking, can only tolerate 50℃ to 60℃ in temperature. Food that are too hot will damage the mucosa, leading to acute esophagitis and acute gastritis.

5. Heavy taste: Malatang is extremely rich and spicy, it also contains fatty ingredients that lead to hyperlipidemia, stomach diseases, duodenal ulcer and other problems.

The doctor continued to tell

美食城堡的安全保卫战
A War to Protect Food Safety

不能一概而论，但是麻辣烫确实存在着一些饮食上的安全问题，你们这些孩子以后要注意到麻辣烫的危害，尽量避免对自己的身体构成不必要的伤害。"大家礼貌道谢后便告别了医生叔叔，连胖胖熊都不禁感慨道："饮食真的是一门大学问，以后吃什么都要弄清楚它的利弊了。雨燕，对不起，我以后会注意的。"雨燕卷着舌头道："没关系的，这也是我们没有听松仔的劝告，以后我们改掉自己不合理的饮食习惯就好了，对吧？"大家达成一致满意而归。

the kids, "Even though we should not generalize the problem, but malatang does contain safety and dietary problems, so it is best to avoid unnecessary harm to your body and pay attention to the malatang that you are eating." Everyone thanked the doctor for his help, Little Bear couldn't help but sigh, "Diet and nutrition are serious science subjects. From now on I need to learn of all the pros and cons before I eat. Sorry, Emily, I will be more careful from now on." Emily replied, "It's okay, we should have listened to Squirrel's warning. From now on, let's change our dietary habits together!" Everyone nodded in mutual agreement.

图9 翻滚的麻辣烫君

Figure 9   Rolling Malatang

美食城堡的安全保卫战
A War to Protect Food Safety

【小微博士有话说】

1. 吃是一门大学问，在饮食的道路上我们要吃到老，学到老。

2. 麻辣烫、大排档等路边小吃安全隐患大，选择的时候一定要谨慎。

[Dr. Micro's notes and tips]

1. Eating is a science, we must learn constantly as we eat.

2. Malatang and outdoor vendors can be a safety hazard, it is best to eat with caution.

美食城堡的安全保卫战
A War to Protect Food Safety

果汁色彩斑斓的秘密
Colorful Secret of Fruit Juice

时值硕果累累的夏天，城堡内各种水果应有尽有，小微博士便给大家留了一个动手任务，周末自己挑选水果为亲人榨一杯新鲜果汁，并要求同学们为自己的果汁拍下美照并附上感想贴到教室后面的心愿墙上。

这天，松仔、胖胖熊还有雨燕约好去超市。大家热情高涨，认真挑选水果、配料。为了使自己果汁更加美味，胖胖熊美滋滋地买了一罐蜂蜜；松仔挑选好橙子后，又称了自己喜欢的核桃、花生等坚果做配料；雨燕只是简单地买了苹果，因为她更喜欢喝白开水。选好食材

It was summer, and Castle City had a variety of delicious fruits. Dr. Micro gave the students a task, to buy some fruits and blend them into juice for their families, and to take some pictures and caption with their thoughts. The photos would be hung on the classroom wall for showcasing.

So, during the weekend, Squirrel, Little Bear and Emily went to the market to buy fruits. They were all excited and picked their fruits rather carefully. To make his juice taste better, Little Bear bought a jar of honey, Squirrel bought some oranges and some walnuts. Emily only bought apples

美食城堡的安全保卫战
A War to Protect Food Safety

后，大家便各回各家准备大展身手。

周一终于来了，可是大家看着并没有想象得那么开心，小微博士提出要看大家照片时，大家都迟迟不愿拿出自己的"作品"。了解情况后，小微博士才知道原来是因为大家都认为自己做的果汁没有超市的美味、好看。小微博士鼓励大家敢于展现自己。雨燕首先拿出了自己的苹果汁照片，结果并不像大家想象的是美丽的青色，而是暗黄色，看着很没有胃口。雨燕说道："我也不知道为什么，我什么东西都没放，做好等妈妈回来后就已经是黄色了，就像坏了一样。"小微博士笑笑示意雨燕坐下，鼓励其他同学展示自己的果汁照片。很多同学

because she liked water more than juice. After selecting their fruits, the three friends all went home to make their fruit juice.

Monday finally came, but it seemed like no one was very happy. Dr. Micro suggested looking at each other's photos, but none of them were willing to share because their juices didn't look as appetizing as the fruit juice sold in the supermarket. After Dr. Micro's encouraging words, Emily stepped up and was the first to show her picture of her apple juice. The picture wasn't the beautiful green color that the students expected, but rather a dark yellowish color, which didn't look appetizing at all. "I don't know why", said Emily "But when my mom came back home to drink the juice, the color changed and

美食城堡的安全保卫战
A War to Protect Food Safety

表示,他们和雨燕一样,很认真在做,没有放错东西做出来却不好喝也不好看。松仔举手说道:"平时在超市买的橙汁特别好喝,我自己做的特别酸,我还放了坚果,做的却没有像超市果汁那么美味,后来我就没有拿给妈妈喝。"胖胖熊窃喜道:"还是我机智,我昨天做的不好喝,我便去超市给妈妈买了一瓶芒果汁,妈妈还夸奖我懂事了。"

教室里炸开了锅,有疑惑,有争论。小微博士示意大家安静并说道:"其实大家已经很棒了,能自己

it looked like as if it were spoiled." Dr. Micro smiled and told Emily to sit down. She then encouraged the others to share their pictures as well. A lot of students shared stories similar to Emily's. They were very diligent in making the juice, but the results were not as tasty or good looking as the fruit juice from the supermarket. Squirrel said, "The tangerine juice from the market were very tasty, but what I made was sour, I even added nuts, but it wasn't very good so I didn't give it to my mom." Little Bear said, "I tried mine and it tasted bad, so I bought a bottle from the supermarket and gave it to my mom."

The classroom erupted with confusion and arguments. Dr. Micro calmed the classroom and said, "You

美食城堡的安全保卫战
A War to Protect Food Safety

动手做果汁就是一个进步，爸爸妈妈会为你们的成长感到开心的。至于为什么你们做的不美味，下面我为大家解答。"小微博士打开幻灯片说道："就像雨燕说的，你们什么都没加，果汁却像坏了一样，其实正是因为你们没有加东西，所以果汁才会坏，超市果汁不会变质的原因是因为它们加了抗氧化剂。"小微博士还告诉大家，水果中含有丰富的维生素，暴露在空气中特别容易被氧化变色从而失去营养价值，所以自制鲜果汁要尽快饮用。大家不解为什么还要榨果汁，直接买不就好了，小微博士耐心地告诉大家，自制果汁更安全、更健康、更营养。至于超市的饮料果汁主要有以下几条危害：

guys all did great! You were able to make fruit juice yourself, I'm sure your parents are all proud of you guys." Dr. Micro opened up a PowerPoint and said, "As Emily said before, you guys did not add anything, but the fruit juice seemed to be spoiled. This is because you guys did not add anything and that is the reason why the fruit juice spoiled. The fruit juice on the supermarket doesn't spoil because they are added anti-oxidation chemicals." Dr. Micro also told everyone that fruits have an abundant number of vitamins, which would lose its nutritious values after exposure to oxygen, that was why self-made fruit juice should be consumed quickly. The students were confused why they should blend fruit juice themselves, and not just bought them instead. Dr. Micro

美食城堡的安全保卫战
A War to Protect Food Safety

1.添加了过多的糖类，热量过高，更易增加脂肪堆积；

2.添加了人工色素以提亮其颜色，这也是果汁色彩斑斓的秘密；

3.维生素破坏严重，且容易变质，影响健康。

雨燕小声喃喃道："还不如直接喝白开水呢？想补充维生素直接吃水果，白开水还能促进新陈代谢、清肠道。"小微博士表示雨燕说得很对，白开水有很多好处，每天早起一杯水，每天八杯水，身体健康肠道好。

explained that self-made fruit juice was safer, healthier and more nutritious. As opposed to the fruit juice from the market, which could have a few negative health effects:

1. Too much sugar, too much calories, leading to accumulation of fat.

2. Added food coloring, this is the reason why fruit juice looks so good.

3. Damaged vitamins, affecting health.

Emily muttered, "Might as well just drink water, and eat fruits to get supply of vitamins. Waters can help metabolism and digestion." Dr. Micro agreed that water has a lot of benefits and that drinking a glass of water in the morning and eight glasses of water a day can make you healthier and have better digestion.

美食城堡的安全保卫战
A War to Protect Food Safety

图 10　白开水更健康

Figure 10　Boiled water is more healthy

【小微博士有话说】

1. 补充维生素等营养物质要注意正确的方法，新鲜水果含量更高。

2. 很多时候越是鲜艳的东西越不健康。

3. 白开水越喝越健康。

[Dr. Micro's notes and tips]

1. Eating fresh fruits is the correct way to supplement vitamins and other nutrition.

2. Often times, the most colorful foods are unhealthy.

3. Drinking water is the way to become healthy.

A War to Protect Food Safety

黑暗料理风波
The Mysterious Dish

美食城堡迎来了一年一度的美食节，在美食节上可以看到各式各样的美味，而且都是全场半价。胖胖熊、松仔和雨燕来到了美食街上，看着每一家所做的美食，有印度飞饼、韩国炒年糕、各式糕点、手抓饼、麻辣小龙虾、水果刨冰等，几个小伙伴都目不暇接，垂涎三尺。突然他们走到了一家黑暗料理店前，被店里的菜色所吸引，就凑上前去看了看。这店里面的菜名让人听着就觉得味道很奇特，像什么西瓜炒香蕉、青菜炒橘子、苦瓜紫薯圆、辣椒蜂蜜青番茄汁、辣椒炒月饼等各种稀奇古怪的菜名，让人听了名字都觉得舌尖发麻。松仔

It was time for Castle City's annual food festival. At the festival, you could find all sorts of delicious foods at half its regular price. Little Bear, Emily, and Squirrel came to the food street, and looked at all the different stands and what they had to offer. Indian Naan, Korean spicy rice cake, assortment of cakes, spicy crawfish, popsicles and other delicious food. The three friends were amazed and began drooling at all the selection. Suddenly, they were strongly attracted by the dishes when they walked into a mysterious restaurant. Some of the dishes had very "interesting" names, such as stir-fry watermelon and

美食城堡的安全保卫战
A War to Protect Food Safety

开玩笑地说："胖胖熊，你不是一向都好吃的嘛，敢不敢尝试一下新花样？"胖胖熊听松仔这么问他，就拍着胸脯说："怎么不敢啦，从出生到现在，除了我不爱吃的东西，就没有我不敢吃的。"说着，就拉着松仔和雨燕走进了这家黑暗料理店。

一进店，胖胖熊就让服务员把菜单拿出来，说真的看到这些菜名，大家真的感觉无法接受。胖胖熊愣了一下，还是硬着头皮点了三份辣椒蜂蜜青番茄汁、橘子泡面和苦瓜紫薯圆。不一会儿，服务员就

banana, stir-fry greens and orange, spicy-honey tomato sauce, stir-fry spicy mooncake and other uncommon names. You would get a weird and tingly feeling just from reading the name. Squirrel said to Little Bear jokingly, "Aren't you always curious about food, Little Bear? Why don't you try these new creations?" Little Bear patted himself on the chest and replied, "Of course! There are no dishes that I am not afraid to try!" Therefore, Little Bear grabbed Squirrel and Emily, and walked straight into this mysterious restaurant.

As soon as they walked in, Little Bear asked the waiter for a menu. Honestly speaking, everyone was rather astonished by the name of these dishes. Little Bear froze for a second, and continued to order three servings of

美食城堡的安全保卫战
A War to Protect Food Safety

把胖胖熊点的东西就送上来了。大家看着桌上的菜都目瞪口呆，没想到竟然真的有人可以把东西做成这样。不过他们还是有点不敢动口吃。胖胖熊鼓了鼓勇气，首先喝了一下点的饮料，感觉酸酸甜甜之外还有一点微辣的感觉，是不同于酸辣汤的另一种口味，但是并不难喝。胖胖熊又大胆地尝试了其他的菜，是那种奇怪的感受，相比较菜肴，胖胖熊更喜欢这种像酸辣汤的味道，竟大口地喝了起来。松仔和雨燕看到后也尝了一下饮料，但还是不能接受这种味道。于是胖胖熊就把给松仔和雨燕点的饮料喝了。付过账后三个小伙伴就走出了饭店。

spicy-honey tomato sauce, tomato ramen, and bitter-gourd yam rings. After a few moments, the waiter brought all the food that Little Bear ordered to their table. The three friends were surprised with their mouth wide open, they couldn't believe that someone would cook up such a thing. They were all too frightened to taste these weird dishes. Little Bear worked up his courage and tasted the drink first, it had a sweet and sour taste to it, but with a little bit of spice as well. The taste of the spice was different from that of sweet and sour soup, but it wasn't bad tasting, so Little Bear decided to try the other food as well. After he tasted the dishes, he thought that he enjoyed the drink a lot more, so he gargled the drink up. Squirrel and Emily followed Little Bear,

A War to Protect Food Safety

and tasted the drink as well, but they couldn't get used to the taste of it. So, Little Bear drank both of their drinks. After they paid their check, the three friends walked out of the restaurant with strange dishes.

As they walked on, Little Bear's stomach suddenly began to ache, and Squirrel immediately realized that maybe Little Bear had gotten food poisoning. Squirrel and Emily sent Little Bear to the hospital right away. After the doctor did a check up with some follow-up questions, he confirmed that Little Bear did indeed get food poisoning. The doctor reminded Little Bear to watch what he had eaten these days, and to eat light meals. Dr. Micro just arrived at the hospital for work, and she bumped into Squirrel. After she learned what had

三个小伙伴走着走着，突然胖胖熊感觉肚子又胀又痛，松仔马上意识到可能刚才胖胖熊吃坏肚子了。松仔和雨燕立刻送胖胖熊到医院。医生询问过病史后，说确实是吃坏肚子了，这两天要多吃点清淡的食物。小微博士正好来医院，碰到了松仔他们，了解情况后，就跟他们说："青西红柿有清热解毒的作用，对化痰止咳是很有疗效的，所以如果平时受寒或者是感冒的人，可以多吃青西红柿，对于肺热咳嗽、喉痛咽干、口舌生疮等有明显的疗效。因为青西红柿没有熟透，所以含有的未成熟的酶也是很多的，这样我们吃进身体，就会影响我们身体酶的活性，对我们的中枢神经也是有影响的，所以应该少量食用。"

just went on, she said to the others, "Green tomatoes can help detoxify the body, it can also help clear phlegm and relieve a cough. Therefore, if you have a cold, you can eat green tomatoes and it will help cure a sore and strip throat. Since green tomato is raw, it contains a lot of enzymes that may disrupt our body's own enzyme activity, also it can affect our central nervous system, so it should be eaten with caution."

图 11 "奇怪"的饮料

Figure 11　Strange drink

美食城堡的安全保卫战

A War to Protect Food Safety

【小微博士有话说】

1. 青西红柿有清热解毒的作用，对化痰止咳也是很有疗效的，所以如果平时受寒或者是感冒的人，可以多吃西红柿，对于肺热咳嗽、喉痛咽干、口舌生疮等也有明显的疗效。

2. 青西红柿没有熟透，所以含有的未成熟的酶也是很多的，这样我们吃进身体，就会影响我们身体酶的活性，对我们的中枢神经也是有影响的，所以应该少量食用。

3. 番茄含有大量可溶性收敛剂等成分，与胃酸发生反应，凝结成不溶解的块状物，容易引起胃肠胀满、疼痛等不适症状，所以空腹的时候最好不要吃。

[Dr. Micro's notes and tips]

1. Green tomatoes can help detoxify the body, it can also help clear phlegm and relieve a cough. Therefore, if you have a cold, you can eat green tomatoes and it will help cure a sore and strep throat.

2. Green tomato is raw where it contains a lot of enzymes that may disrupt our body's own enzyme activity, also it can affect our central nervous system, so it should be eaten with caution.

3. Tomatoes contain large amounts of soluble astringent ingredients, and can react to our body's stomach acid, causing condensed lumps that cannot be dissolved. This may lead to gastrointestinal fullness, stomach ache and other conditions. Therefore, it is best not to eat tomatoes on an empty stomach.

美食城堡的安全保卫战
A War to Protect Food Safety

黑心小作坊
The Dishonest Shop

在一个阳光明媚的清晨，伴随着一阵阵"丁零零……"闹钟声响，熊妈妈起床去做好早饭，并嘱咐熊爸爸相关的事宜后就提着行李出门了。熊爸爸看熊妈妈出门后就把胖胖熊叫起来了一起洗漱完后去吃早饭。熊爸爸对胖胖熊说："妈妈去郊游了，这两天由爸爸照顾你，你想吃什么，爸爸中午给你做。"听到这话，胖胖熊用怀疑的眼神看着熊爸爸问道："爸爸，你真的可以吗？"熊爸爸拍拍胸脯说："宝贝，你要相信爸爸，爸爸之前只是深藏不露，没有展示自己精湛的厨艺，今天爸爸就露一手给你瞧瞧。一会儿你去找松仔玩会儿，中午叫松仔

"Bzzzzzz…", the alarm clock had set off on this lovely morning. Mother Bear woke up, made breakfast, and left the house with her luggage on a trip after reminding Papa Bear of a few things to take care of while she was gone. After Mother Bear left, Papa bear went to wake Little Bear up for breakfast. "Your mother went on a fieldtrip, so I will be taking care of you these days. What would you like to eat for lunch? I will cook for you!" said Papa Bear. Little Bear looked at his father in askance, "Are you sure you can cook?" he asked. Papa pumped his chest with confidence and said "Of course, I can! I was just

美食城堡的安全保卫战
A War to Protect Food Safety

一起过来吃饭，让他也尝尝你老爸我的厨艺。"胖胖熊只好相信了爸爸，于是吃过早饭就跑去找松仔玩了。

熊爸爸看时间差不多就在厨房开始忙碌起来，不过毫无疑问地，厨房被熊爸爸弄得一片狼藉。熊爸爸心想，可一定不能让胖胖熊失望啊。于是灵机一动，可以定份外卖送过来呀，让胖胖熊以为是我做的不就可以了。于是他在网上订了几个胖胖熊喜欢吃的菜，没一会外卖就送过来了，熊爸爸将菜盛在盘子里，等胖胖熊和公仔回来吃饭。不久，胖胖熊带着松仔回到家，看到

hiding my skills before, now it is time for me to show them off. Why don't you go play with Squirrel for a bit, and when I've done cooking, you can invite him to have lunch with us too! Both of you can have a try at my delicious cooking!" Little Bear could only trust Papa Bear, and after breakfast he went to find Squirrel.

Papa Bear looked at the clock, it is time to get busy! But, just as we thought, Papa Bear messed up the entire kitchen attempting to cook up a meal as promised. But Papa Bear didn't want to disappoint Little Bear, since he promised his son a meal, and he ordered takeout instead, hoping that Little Bear would think it was his own cooking. Papa Bear ordered a few of Little Bear's favorite dishes online, and the takeout

美食城堡的安全保卫战
A War to Protect Food Safety

桌上丰盛的午餐,口水立马就流了出来。

三个人没一会儿就把桌上的菜都吃完了。这一餐可把胖胖熊乐坏了,一直吵着以后都要吃熊爸爸做的菜。但是接下来的几天就苦了熊爸爸了,他每天都要装作大厨的模样在厨房忙碌,背地里却偷偷订外卖。接下去的那两天,胖胖熊吃得很是开心,但结果就在第二天晚饭后突然出现恶心呕吐的症状,这下可把熊爸爸急坏了,赶紧和松仔

meal was delivered to the doorsteps in no time. Papa Bear poured all the food out of its plastic delivery containers into silverwares they had, set the table, and waited for Little Bear and his friend Squirrel to come home for lunch. Shortly afterwards, Little Bear arrived from the playground with Squirrel. They saw the feast on the table, and started to drool immediately.

The three of them finished all the dishes in no time. Little Bear was so happy, because this meal was outstanding. From now on, Little Bear wanted to eat Papa Bear's home cooked meal every day! The next couple of days were very busy for Papa Bear, he had to pretend to be a chef in the kitchen and pretend to cook, then secretly order food from restaurants online. Little Bear on the other had

A War to Protect Food Safety

把胖胖熊送到了医院。结果医生说是由于食物中毒引起的。松仔马上说："这几天胖胖熊都是吃熊爸爸做的饭怎么可能会食物中毒呢？"。熊爸爸惭愧地说："真不好意思，这几天你们吃的饭都是我订的外卖，应该是外面的饭菜不干净导致的。"松仔听完后说："熊爸爸，如果真是外卖的问题，你应该报警啊，不然以后还会有人出现这样的情况。"在一旁看到这一幕的小微博士走了过来，对熊爸爸说："你把今晚吃的食物拿过来，我化验一下，看到底是不是你所订的外卖导致食物中毒的。"

been enjoying the "home cooked" meals a lot. However, after dinner on the second day, Little Bear's stomach began to feel very uncomfortable and it made him feel nauseated. Papa Bear was very worried. He and Squirrel quickly sent Little Bear to the hospital. After the check-up, the doctor said that Little Bear's condition was caused by food poisoning. "How can that possibly be? Little Bear has been eating home cooked meals made by Papa Bear!" Squirrel exclaimed with confusion. Papa Bear blushed with embarrassment and said, "I am so sorry, but the truth is that I have been ordering delivery food online instead of cooking it myself. The restaurant that I bought the food from was probably unclean." After hearing Papa Bear's confession, Squirrel said, "If there is

美食城堡的安全保卫战
A War to Protect Food Safety

something wrong with the food being delivered, you should call the police and report the restaurant! Or else they will keep selling the harmful food to other people and cause more people to become sick!" Dr. Micro had been standing nearby, she saw the whole thing happen and said to Papa Bear "Why don't you bring me the food you guys ate today, let me examine it to see whether it was the cause of the food poisoning."

熊爸爸就回家将剩下的饭菜带到了小微博士那里，经过一个多小时的化验，小微博士对熊爸爸和松仔说："确实是这些食物导致的，由于胖胖熊吃的比较多，体内毒素积聚多因而导致食物中毒。你们吃的比较少，毒素轻，可以靠自身的防御机制排出体外，但也不能再吃这种食物啦。食物中毒是由于食材

Papa Bear went home and brought all the leftover delivery to Dr. Micro's lab, after an hour of analysis, Dr. Micro said to Papa Bear and Squirrel, "Little Bear's food poisoning was in fact caused by the delivery. Little Bear had a lot to eat, so the poisonous substances gathered slowly eventually causing food poisoning. You guys didn't eat as much

美食城堡的安全保卫战
A War to Protect Food Safety

不新鲜、烹饪不卫生导致的。像这样的情况最好像松仔所说的那样报警，以免更多的人受害。"

熊爸爸听了后就打电话报了警。警察经过侦查，发现熊爸爸订外卖的这家店，放在网上的照片是菜品色泽艳丽，厨房不锈钢灶具洁净透亮。而实体店的厨房却是昏暗狭小的制作间，墙上、灶台上、饭锅上到处是黑乎乎的油渍；老板从外面买来的火腿肠，用牙咬开外包装就直接分切配到炒饭中；掉进脏东西的饭盒，在桌上磕打一下，就

as Little Bear so it didn't affect you all as much as Little Bear. Although your body is able to defend itself from such harm, it is best if you stop eating from the delivery. Food poisoning is caused by eating raw foods, or cooking food in an unsanitary environment. You should report the restaurant to the police as Squirrel suggested, so that other people won't be harmed."

Papa Bear called the police immediately. After the police conducted their investigation, they realized that the restaurant that Papa Bear ordered from displayed delicious photos of great dishes on their website, but the restaurant's kitchen is actually very dirty with oil and trash all over. The owner would rip open the sausages he bought with his teeth and put it straight

美食城堡的安全保卫战
A War to Protect Food Safety

直接装饭；用完盛饭板直接放在全是污渍的锅盖上。熊爸爸和松仔听到这一消息惊呆了，没想到自己会遇到这样的事，心想以后还是在家里吃饭为妙。

into the frying pan. If something dirty went into the food containers, the restaurant would just "empty" it by hitting it on the table, and put food in without washing it. After hearing this, Papa Bear and Squirrel were astonished, they couldn't believe their ears, and they decided to cook at home from now on.

图 12　小作坊脏乱差

Figure 12　Dirty and messy small workshop

美食城堡的安全保卫战

A War to Protect Food Safety

【小微博士有话说】

1. 对于网上订餐，最好选择有卫生合格证的实体店订餐，杜绝黑心小作坊。

2. 无论外面的东西有多好吃，自己做饭吃更健康。

[Dr. Micro's notes and tips]

1. When ordering online, it is best to select actual restaurants with a storefront and certificate. Please avoid these dishonest and unsanitary shops.

2. No matter how good the food is in a restaurant, food cooked at home always remain the healthiest.

美食城堡的安全保卫战
A War to Protect Food Safety

胡萝卜让你眼睛美又亮
The Beautifying Carrot

树儿绿了，花儿开了，这天美食城堡学校举办了一年一度的春游活动，同学们都带着各自的午餐兴高采烈地出发了。经过一个小时的整理场地，同学们开始游戏了，玩得很是愉快，不知不觉到了中午，小微博士组织同学们围成一圈一起吃饭。松仔看到松鼠妈妈为自己准备的是胡萝卜不由得皱起眉头，因为胡萝卜是松仔不喜欢的蔬菜，雨燕看到松仔迟迟不下筷子就问松仔怎么了。

It was spring time and the leaves on the tree had turned green, and the flower blossomed. Castle Elementary School organized its annual spring fieldtrip! All the students have prepped their lunches and they're ready to go. They arrived at the outing site, and after an hour of preparing the space, everyone was ready to play! Time flies by when you're having fun, it was already lunch time. Dr. Micro gathered all the students around a circle to have lunch together. Squirrel opened up his lunch box and he was rather upset, because his mother had prepared carrots for him, and carrots were his least

松仔不开心地回答道:"我很不喜欢吃胡萝卜,妈妈为我准备的是胡萝卜炒腊肉。"

小微博士突然再次提醒同学们下午他们要志愿捡垃圾、打扫草坪,所以会比较累,建议大家一定要吃得饱饱的。松仔走向小微博士要求去附近买点面包,小微博士问完原因后让松仔坐回了原位,接着他便请不喜欢吃自己带食物的同学举手示意。松仔和胖胖熊难过地举起手来,原来熊大妈也给胖胖熊准备了他不喜欢吃的胡萝卜。

favorite vegetable. Emily saw Squirrel frowning at his lunchbox and asked "What is wrong, Squirrel?"

"I dislike eating carrots, but my mom cooked some stir-fry carrots and sausages," answered Squirrel with disappointment.

All the other students had dug into their lunches, so Dr. Micro reminded all the students to eat plenty, because in the afternoon they would be volunteering to clean up the grass field, and it would be a tiring task. Squirrel walked to Dr. Micro and asked if he could have permission to buy some bread from a shop nearby. Dr. Micro asked for his reasons and then told him to sit back to his seat. Next, Dr. Micro asked all the students to raise their hands if they dislike the food they had for lunch. Both

小微博士立马以"胡萝卜知识知多少"为题举行了有奖竞答赛,大家你一言我一语,你说胡萝卜是蔬菜,吃了长得高,他说胡萝卜可以补充能量供我们走路。聪明的雨燕总结了大家的观点,又告诉大家胡萝卜能让我们的眼睛更加明亮,建议大家一定要多吃胡萝卜。小微博士鼓掌以示自己的赞同,同时拿出了微型电子投影仪,将胡萝卜的资料投影到绿油油的草坪上。

投影资料显示:胡萝卜被誉为"东方小人参",它的作用如下:

Squirrel and Little Bear raised their hands. It had turn out, Mother Bear also prepared carrots as lunch for Little Bear.

Dr. Micro immediately started a Carrot Q&A contest, everyone joined to discuss the nutritional value of carrots. Some suggested that eating carrots can provide us with energy so we can walk more, others said that carrots are a type of vegetable, therefore, it helps kids to grow taller. The smart Emily brought everyone's ideas together and added that eating carrots can help improve our eyesight, so it is best to eat more. Dr. Micro applauded everyone's suggestions, she then brought out a projector and projected some information on carrots onto the grass field.

The information displayed described carrots as "The Eastern Ginseng", with

A War to Protect Food Safety

1. 防止血管硬化，减少心脏病；

2. 胡萝卜素能转变成维生素A，增强机体免疫力；

3. 胡萝卜素和维生素A能促进眼内感光色素生成能力，预防夜盲症，减缓眼睛疲劳，益肝明目；

4. 胡萝卜中的植物纤维有很强的通便能力；

5. 其中的木质素具有提高巨噬细胞（保卫人体健康的士兵）的能力，预防感冒等作用；

6. β-胡萝卜素还能有效地预防花粉过敏现象。

小微博士关掉投影仪说道："胡萝卜功效如此之多，我们是不是

the following effects:

1. Prevent arteries from hardening and reducing heart diseases;

2. Carotene in the carrots can be converted into Vitamin A, which enhances our body's immunity;

3. Carotene and Vitamin A can help the formation of photopigment in our eyes, which reduces eye fatigue and nyctalopia;

4. Carrots contain a lot of plant fiber which helps with our bowel movement;

5. Xylogen enhances the macrophage (protecting people's health) to prevent colds, illnesses and so on;

6. Carotene B is also effective in preventing pollen allergy and other allergic reactions.

Dr. Micro shut off the projector and said "Carrots are very effective

美食城堡的安全保卫战
A War to Protect Food Safety

应该多吃胡萝卜增强自身健康呢？最重要的是，我们要养成不挑食的好习惯，因为我们的身体需要各种营养物质，我们的膳食应该满足我们的身体需求。"

听完后松仔和胖胖熊默默地拿起了眼前的饭盒，下午大家能量满满地做完了志愿捡垃圾活动，林场管理员还夸同学们并授予每位同学"志愿小能手"徽章。回校的路上，小微博士给每个人留了一个任务，让父母选择一种蔬菜，同学们查阅资料找到这些蔬菜的作用。

in improving our health, so shouldn't we eat more carrots to make ourselves stronger? The most important thing is that we must not be picky with our foods, because our bodies need all kinds of nutrients, so having balanced meals is rather important to support our body's needs."

After listening to Dr. Micro's presentation, Squirrel and Little Bear picked up their lunchboxes and ate the vegetables. In the afternoon, all the students were full of energy and completed their volunteer work of picking up trash in the forest. The forest keeper also commended the students for their services, and presented each of them a volunteer badge for appreciation. On the way back home, Dr. Micro gave each student an assignment: to let

美食城堡的安全保卫战
A War to Protect Food Safety

次日课堂上，同学们都积极参与，踊跃发言，说出了自己的发现，并表示自己以后一定多吃各种各样的蔬菜，健康成长。小微博士对大家的表现很满意，这个活动也受到了家长们的一致好评。这天下午回到家中，松仔主动提出要吃妈妈做的胡萝卜炒腊肉，同时他也发现胡萝卜不仅没有那么难吃，还很香甜。

their parents pick a kind of vegetable, and they will conduct research on the selected vegetable.

The next day in class, all the students participated in the discussion on vegetables. They each presented their discoveries on their selected vegetable, and now everyone learned more and were keener on eating vegetables to maintain a healthy body. Dr. Micro was pleased with everyone's assignment, and the parents also praised this assignment as well. After school, Squirrel went home and suggested to Squirrel mother to make stir-fry carrot and sausages, he also found that maybe carrots were not so bad tasting after all, it actually had a great taste to it!

美食城堡的安全保卫战
A War to Protect Food Safety

图 13 抱着胡萝卜的兔子

Figure 13　A rabbit hugging a carrot

【小微博士有话说】

1. 蔬菜有很多作用，我们应该平衡饮食不挑食。

2. 任何食物都有它的作用，我们都要吃，但不能多吃，也不能少吃，均衡饮食，健康生活。

[Dr. Micro's notes and tips]

1. Vegetables have many effects, so we should not pick our favorites, a balanced diet is key to our health.

2. Each vegetable has its own special effects, so we must eat them all. However, we must not eat too many or too little of each. Remember to always maintain a balanced diet.

美食城堡的安全保卫战
A War to Protect Food Safety

化妆的鸡蛋
An Egg with Makeup

春天是个万物生长的季节，一大早熊妈妈就在寻思着，"吃点儿什么好呢？一年之计在于春，可不能让胖胖熊错过长个头的好时节。"突然，熊妈妈想起去年也正好是这个时候，邻居家的小宝吃了好些鸡蛋，结果就像开了挂似的一整年里猛长个儿呢。她兴奋地两手一拍，如同得到了通关秘籍，说道："对，就是它了！"

嘿，还真别说，鸡蛋在美食城堡的超市里人气可高啦。且不算风格各异的包装，光是各种名字：五

Spring is the season for growth. It was an early morning, and Mother Bear was thinking of what to cook, "What is good to eat? Spring is the best season of the year, I cannot let Little Bear miss out on the season for growth!" thought Mother Bear. Suddenly, Mother Bear recalled that their neighbor had eggs during this time last year, and their child grew really fast after that. Mother Bear clapped her hands with excitement as if she had just found a secret treasure. "Oh yes, eggs it is!"

To Mother Bear's surprise, eggs sold in the Castle City Super Market were extremely popular. They had all sorts

美食城堡的安全保卫战
A War to Protect Food Safety

谷蛋、虫草蛋、柴鸡蛋、土鸡蛋、红心蛋……就让熊妈妈犯了难，于是她决定先去找小微博士做做功课。

小微博士推了推眼镜，说道："其实这是部分商家的一种营销手段。他们借用了土鸡蛋在消费者心中的'绿色概念'，趁机炒作，使消费者对土鸡蛋的认知变得模糊。"

"那究竟什么样的鸡蛋才是真正的土鸡蛋？"小微博士决定带着胖胖熊一家，一同去走访美食城堡里的散养鸡养殖场。

据该基地的老板介绍，他承包了十亩林地，专门用于养鸡。只见偌大的山上，植被茂密，溪水潺流，三五成群的鸡在欢快地奔跑

of different packaging and names such as whole eggs, corn eggs, caterpillar eggs, free-range eggs, red yolk eggs and many more. Mother Bear had a very hard time trying to make a decision, so she went to Dr. Micro for some suggestions.

Dr. Micro pushed his classes and said "Actually, this is a common type marketing strategy. They use the "organic" and "green" concept and market it to the customers to hype up the market."

"What are real free-range eggs then?" Dr. Micro decided to take Little Bear and his family to visit Castle City's chicken farm to find out.

According to the farm owner, he has contracted over 6000 square meters of land just to raise chickens. It can be seen mountains, streams, and a lot of

A War to Protect Food Safety

着，它们可以刨土、吃草、啄虫，也可以定时吃到老板从村里收购来的玉米、小麦等谷物。就这样在林地里自然地成长，这才是真正意义上的土鸡，它们产的蛋才真是名副其实的土鸡蛋。

对于超市上出现的各种鸡蛋，养殖场老板笑道，"这还真需要一双火眼金睛呢。如果鸡这两天吃的是虫子，那它产的就是虫草蛋；过几天再喂它谷物，产的就五谷蛋。那我出售的鸡蛋到底是该称虫草蛋，还是五谷蛋，还是土鸡蛋好呢？所以啊，这些都是商家炒作出来的'概念蛋'。"

养殖场老板还坦言："价格低于

vegetation on the farm. Several groups of chickens are happily running around the farm. They can eat the grass, peck and eat the bugs, and they can also eat chicken feed from the farmer. Only chickens in such an environment can be called a "free-range chicken" and lay authentic "free-range eggs".

The farm owner smiled and answered the questions regarding all the different type of eggs on the market, "You'll need to take a closer look. If your chickens are eating bugs, then they can be called "caterpillar eggs", if they eat grains then they are "grain eggs". So, what would you call my eggs then, since I feed my chicken eating everything? Should it be called grain eggs or free-range eggs? It is all a marketing strategy."

The farmer then explained, "If the

15元一斤的土鸡蛋要慎买。"他是这样解释的：真正的土鸡一般一天只产一个蛋，而且并不会连着天天下蛋，再加上包装、运输、转卖等环节的费用，商家至少要卖到15元每斤才能勉强保本。

他一边说着一边给小微博士看刚从产蛋房里收回的土鸡蛋。这些鸡蛋的表面有层薄薄的白霜及细密的麻点，大小不像超市里的那样匀称，颜色也是不尽相同。因为每一只鸡在散养过程中，多少都会受到环境等多方面的影响，且鸡本身也存在着个体差异，所以产的蛋在外观上自然也就不同。

至于蛋黄的颜色，养殖场老板

price of the free-range eggs is under 15 RMB per 500g, then do not consider buying them. Most free-range chickens can only lay an egg a day, and they cannot lay an egg every day. If you add in the cost of packaging, transportation, and sales, the store must sell the eggs for at least 15 RMB for every 500g, just to break even."

As he introduced the eggs, the farmer showed Dr. Micro the free-range eggs that were just collected from the farm. The eggs have a thin layer of white chalk and black dots all over, they do not look like the clean and even eggs at the supermarket at all, even the colors are all different. That is because each chicken has a different lifestyle on the farm, so their bodies and eggs produced are different as well.

As to the color of the yolk, the

说，"这与鸡的食谱很有关系，比如吃玉米就会偏红，吃稻谷、青菜等就会偏黄。真正的散养鸡很难保证每一枚鸡蛋的颜色都统一，毕竟每一只鸡的习性都会存在差异，而只有通过专门的饲料喂养或加了某种色素，这些蛋黄颜色才可能会成统一的深黄或橘红色。"

听养殖场老板一说，熊妈妈恍然大悟，超市里那些"体型一样娇小、蛋壳白净、蛋黄颜色越深，鸡蛋质量就越好"的说法，还真是有点儿不靠谱呢。

farmer further explained to Mother Bear "This has to do with the chicken's meals, for example, if they eat corn then the egg yolk will appear red. If they eat grains or vegetables, then the yolk will most likely turn yellow. You cannot ensure that every single free-range chicken egg is the exact same color or size, because each chicken is different. Only chickens that are fed chicken feed and raised in a non-free-range chicken farm will have a standardized color and size."

Mother Bear finally knew the truth after she had listened to the chicken farmer's explanations. People say that high quality eggs have white shells, darker colored yolk, and that they are usually smaller in size, but now, it seems that the saying is not reliable at all.

A War to Protect Food Safety

图14 各种各样的鸡蛋

Figure 14　All kinds of eggs

【小微博士有话说】

人们普遍认为，散养的土鸡每天在林地里活动觅食、吃菜、啄虫等，虽然容易因为营养不均衡，导致下的蛋往往个头比较小，但因为食材天然，蛋黄中的类胡萝卜素和维生素B2含量较高，因此蛋黄会更大、颜色更深一点。而"洋鸡蛋"因单一采用了人工饲料，其营养价值可能不如土鸡蛋。但事实上，现在科学配比的饲料也可以囊括多种营养物质，而且土鸡蛋的胆固醇含

[Dr. Micro's notes and tips]

Most people believe that free-range chickens are not nutritious and that they produce smaller eggs, since they run around in the wild and peck on bugs and vegetables. However, because of the natural ingredient the chicken eats, free-range eggs usually contain a higher concentration of Vitamin B2 and carotene, so the yolk is larger and darker in color. On the other hand, chickens raised in a hennery are being

美食城堡的安全保卫战
A War to Protect Food Safety

量普遍比洋鸡蛋高出2~3倍，并不适合老年人长期使用。因此撇开不良商家的有害化工添加剂，就营养价值来说，土鸡蛋和洋鸡蛋各有千秋。但就口感而言，确实是土鸡蛋更胜一筹。

那该如何辨别土鸡蛋呢？

1.看蛋壳。商家虽然会按鸡蛋的大小分类，但在同一批大小的鸡蛋中，土鸡蛋的确是会存在较大的

fed chicken feed, therefore the eggs are not as nutritious as free-range eggs. In reality, scientists have found that the chicken feed now contain a variety of nutrients, and that the cholesterol content in free-range eggs are generally 2 to 3 times higher than normal eggs. It is not suitable for seniors to consume over a long period of time. If we put aside the farms that adds harmful ingredients into the animal feed, and compare free-range and normal eggs by itself, they each have their own values and advantages. However, in terms of taste, free-range chicken may win.

How to differentiate free-range eggs?

1.Observe the shell. Even though the store will sort the eggs by size, free-range eggs will appear significantly

105

美食城堡的安全保卫战
A War to Protect Food Safety

差异。因此可通过同一批鸡蛋颜色差异的比较，作出基本判断。

2. 看蛋清。土鸡蛋蛋清黏稠，而且特别透亮，腥味很淡。

3. 看蛋黄。土鸡蛋蛋黄硕大，且因脂肪含量高，黏度也高，不易打散；煮熟后口感圆润微甜，非常细腻，吃后口齿留香。

此外，小微博士友情提示。生吃鸡蛋不但不利于营养的吸收，久食会增加肝脏负担，还易患上肠胃炎哦！

different than other eggs. Therefore, you can tell by the color of the shell whether the eggs are free-range or not.

2. Check the egg-white. Free-range eggs have a light, clear, and almost taste-less egg whites.

3. Compare the yolk. Free-range eggs generally have a very large yolk. Since the fat content is rather high, it is hard to stir up its yolk. After boiling the egg, the yolk is very tender and soft.

Dr. Micro also reminds everyone that eating raw egg is not only bad for you, but it can also damage your liver and spleen in the long-run, and you may even get gastroenteritis.

美食城堡的安全保卫战
A War to Protect Food Safety

皮蛋皮蛋是坏蛋
Preserved Egg is a Bad Egg

冬雪初融，寒气未消，美食城堡的人们就迎来了一年一度的狂欢节——春节。城堡的大街上人来人往，有买年画的，有买年货的，有买玩具的，还有穿上新衣服跑来跑去玩的孩子们，好一派喜庆的景象。

春节是中国最古老的节日，是一年四季中最隆重的日子，而团圆饭则是其中最重要的时刻。大年三十晚上，家家户户都要聚在一起吃团圆饭，餐桌上摆满了各种各样的美食。

Winter is almost over, but it is still a little chilly outside. Castle City's annual festival celebration has arrived—Chinese New Year! Many people are on the street right before the holiday, and they buy all sorts of goods such as food, toys, festive paintings, new clothes and others. The kids are also running around with joy. What a festive scene.

Chinese New Year is China's oldest holiday, and it is also the grandest celebration in the entire year. Eating a family reunion meal during Chinese New Year is the most important moment of the holiday. The night before

美食城堡的安全保卫战

A War to Protect Food Safety

一大早熊妈妈就出门去采购年货了，胖胖熊可高兴了，一想到过年了又能吃好多好美味的东西，他就开心得不行，一边的熊爸爸则在认真地打扫卫生。不久，熊妈妈就买完年货回来了，买了不少，幸亏隔壁的熊大叔帮着扛回来，不然她都搬不回来。熊妈妈连连跟熊大叔道谢，然后就走进厨房开始准备年夜饭了，她每年都要做好多个菜，今年当然也不能例外。熊爸爸时不时地进去帮一下熊妈妈，胖胖熊则在客厅看着电视，他只需要乖乖等晚饭就好。

Chinese New Year, everyone will eat a delicious meal with a lot of wonderful dishes with their family members.

Mother Bear went out early in the morning to buy goods for the holiday. Little Bear was so happy because he got to eat so many tasty food during the celebration, just thinking about it made him giggle. Papa Bear on the other hand was cleaning the house to prepare for the festivities. Soon, Mother Bear came back from the market, and she bought so much stuff that their neighbor Mr. Black Bear had to help carry the things into their house. Mother Bear thanked Mr. Black Bear for his help, and then went straight into the kitchen and began to prepare for the reunion dinner. She makes plenty of foods every year, and this year is no exception. Papa Bear would pop in

A War to Protect Food Safety

时间过得很快，胖胖熊开始饿了，他开始跑去催熊妈妈，刚好熊妈妈已经在准备最后一道菜了，是胖胖熊喜欢的皮蛋，他在一旁叮嘱妈妈多做一点，熊妈妈说："做多了吃不完，快去准备一下，我们要开饭了。"不一会儿，丰盛的年夜饭便摆满了一桌，胖胖熊一家围坐桌旁，共吃团圆饭，满桌的佳肴美味，满屋的快乐气氛。

the kitchen to help Mother Bear with cooking from time to time, and Little Bear was just sitting in the living room, watching TV, and waiting for the food to be prepared.

Little Bear got hungry, so he went into the kitchen to urge his mother to cook faster. Mother Bear was just finishing the last dish, it was Little Bear's favorite preserved eggs, and Little Bear told her to make more because he was rather hungry, "We won't be able to eat that much food, it is almost ready, go and prepare the table" said Mother Bear. Finally, the delicious feast was ready, and the table was filled with food. Everyone sat around the table and began to feast on the New Year reunion meal. Everyone was having a great time, and the room was filled with

美食城堡的安全保卫战
A War to Protect Food Safety

过了大年三十,接下来春节的一个重要环节就是拜年。大年初一,美食城堡的小伙伴们就开始拜年了。胖熊熊去了离家不远处的松仔家,然后他又开始犯懒病了,不想走了,说要回去睡懒觉。松仔连忙拦住他,并说:"大过年的,睡什么觉,我们一起去雨燕家拜年吧。"在松仔的劝阻下,胖胖熊没有回家睡觉而是去了雨燕家。去过雨燕家后,雨燕提议大家一起去给小微博士拜年,小微博士平时帮了大家不少忙,美食城堡的伙伴们都很喜欢她,大家就都一起去了。小伙伴们带了很多礼物,小微博士特别高兴,热情地款待大家。大家一起开心地聊着天,小微博士问大家年夜饭都吃了哪些美食,大家争先恐后地回答着,只有胖胖熊在一旁

joyous laughter.

The reunion dinner was great, and the next morning came the important holiday ritual of greeting friends and family. On the day of Chinese New Year, all the citizens of Castle City are out greeting their loved ones. Little Bear went to Squirrel's house because he lived close by, afterwards he got very lazy and wanted to return home for a nap. "Come on, it's New Years, let's go greet Emily's family!" Squirrel said, and they went to Emily's house. They arrived at Emily's, and Emily suggested that everyone should go greet Dr. Micro, as she's been very supportive of the city, everyone in the city loves him. The three friends brought Dr. Micro many gifts. Dr. Micro was very happy to see everyone and she invited everyone

A War to Protect Food Safety

没精打采，小微博士见状，特地走过来问胖胖熊："胖胖熊，年夜饭吃了啥？"胖胖熊听小微博士问，开始慢吞吞地回答："红烧排骨、年糕、皮蛋……"，刚说完，一直在一旁认真听着的雨燕发话了："皮蛋不是好东西，不能吃。"胖胖熊这下精神来了，立马反驳道："皮蛋怎么不好了，怎么就不能吃了？"小微博士见两人这阵势似乎要吵起来，连忙打断说："雨燕说得不是没有道理，可能大家还不是很了解皮蛋，那我就给大家普及一下吧，注意认真听喽，听完我请大家吃点心。"

for snacks. They all began to chat with enjoyment. Dr. Micro then asked what everyone had for reunion dinner, and they all happily introduced the dishes at their table. Little Bear on the other hand was silent, Dr. Micro saw Little Bear all alone and came forward to ask him, "What did you have for dinner Little Bear?" Little Bear greeted Dr. Micro and began to reply slowly, "Beef, rice-cake, preserved eggs…" just as he finished his sentence, Emily came and said, "Preserved eggs are bad for you, you shouldn't eat it." Little Bear was startled and immediately fought back, "What do you mean? Preserved eggs are delicious!" Dr. Micro saw that the two was about to start an argument, so she quickly added, "Emily is right to some degree, but it's best for everyone to

美食城堡的安全保卫战
A War to Protect Food Safety

总的归纳为以下几点，听我妮妮道来哦。

1. 蛋白质变质

很多人在平时生活中都非常喜欢吃皮蛋瘦肉粥、凉拌皮蛋，适当的食用的确十分美味，但如果过量食用的话对人体的危害却是非常严重的。很多人都不知道皮蛋也会影响到人体健康，专家提醒皮蛋这种食物少吃无妨，却不能过量食用。这是由于皮蛋的腌制过程十分复杂，通常都是用茶叶、石灰泥包裹制成的，因此在制作过程中就难免会使用大量的儿茶酚、鞣酸和氢氧化钠，这些物质会侵入蛋体的蛋白质中，从而导致蛋白质分解、变质，因此腌制成功的皮蛋中不仅营

learn more about preserved eggs first. Let me tell you about them, and then afterwards we can all have a snack."

Dr. Micro summed it up in the following points.

1. Protein deterioration

Preserved egg is a rather common type of food, everyone loves them and they are put in congee and made as a cold dish. It is very delicious, but if you eat too much then it could become harmful. Most people don't know that preserved eggs may affect your health. Health professionals suggest that eating a few is fine, but do not go overboard. The process of making preserved eggs is rather complex, it is usually made by wrapping the egg with tea or lime plaster. Therefore, large amounts of substances such as catechol, tannin,

养价值遭到了破坏，同时大量食用还有可能会过量摄取变质的蛋白质。

2. 铅中毒

皮蛋虽然美味，但其中所含有的营养少，除此之外在皮蛋中还含有一定量的铅。如果在平时生活中过量食用的话，就会导致铅中毒的情况。因此很多人在购买皮蛋的时候都会选择一些无铅皮蛋，然而专家提醒无铅皮蛋可能同样含铅，只是其中的铅含量比较少而已。因此这种皮蛋成年人适当食用还行，对于儿童还是少吃为好。

3. 影响智力发育

皮蛋对于儿童而言是十分不健康的。首先，铅一旦进入儿童体内

and sodium hydrate will enter the egg. This leads to protein decomposition, deterioration, so successfully made preserved eggs' nutritional value has been damaged.

2. Lead Poisoning

Although preserved eggs are delicious, it contains nutrition. It also contains less lead. If you eat an excessive amount daily, then it may lead to lead poisoning from lead intake. Therefore, many people will choose to purchase lead free preserved eggs, however, experts remind us that lead-free eggs may also contain a small amount of lead. Therefore, preserved eggs may only be suitable for adults to eat, but children should eat less of it.

3. Affects intellectual development

Preserved egg is very unhealthy

美食城堡的安全保卫战
A War to Protect Food Safety

就很容易存留在肝、肺、肾、脑等组织及红细胞中，长此以往势必导致牙齿以及骨骼中钙的流失，从而严重地影响儿童的正常骨骼生长以及牙齿生长。其次，皮蛋中的铅还有可能会导致孩子出现侏儒现象。最后，铅还有可能会导致儿童出现发育不良、食欲减退、胃肠炎等病症，严重的话还有可能会影响儿童的智力发育。

4. 食物中毒

很多人都喜欢在炎热的夏季傍晚吃一份凉拌的皮蛋，但是专家提醒，这个时候吃皮蛋稍有不慎就有可能会导致食物中毒。还有很多人喜欢在饮啤酒时用皮蛋助兴，研究中发现干净的皮蛋蛋壳上只有

for children. First of all, once the lead enters the children's body, it can easily remain in the liver, lung, kidney, brain tissue and red blood cells. After a long time, it will lead to inhibition of teeth growth and loss of calcium in the bone, thus it may seriously affect children's health and their normal body development and growth. Lead in the preserved eggs may even cause dwarfism in children. Lastly, lead can also cause stunted growth, anorexia, gastroenteritis, and in some serious cases, affect the mental development in children.

4. Food poisoning

A lot of people likes to enjoy a plate of cold preserved eggs on a hot summer night, but experts remind everyone that eating preserved eggs may lead to food poisoning. Many

A War to Protect Food Safety

400~500 个细菌，而一些脏的皮蛋蛋壳上则有高达 1.4 亿~4 亿个细菌。这些细菌若大量通过蛋壳的孔隙进入蛋内，食用之后便会导致食物中毒。

总之，皮蛋利弊并存，根据个人喜好，适当食用为宜。皮蛋不宜存放冰箱，不宜与甲鱼、李子、红糖同食。皮蛋虽然好吃，但存在有大量的健康隐患，因此在平时生活中应该少吃些皮蛋，尤其是处于生长发育期间的儿童，更要注意少吃。

people love to eat preserved eggs while drinking beer. A study has found that the cleaned preserved egg shell contains about 400 to 500 bacteria, but some dirty preserved egg shell can contain as high as 140 million to 400 million bacteria. The large number of bacteria can enter the egg through the pores of the eggshell, after eating it will cause food poisoning.

All in all, there are pros and cons to preserved eggs, and you can eat an appropriate amount depending on your personal preference. Preserved eggs should not be stored in the refrigerator, also, it shouldn't be eaten together with soft shelled turtle, plum and brown sugar. Although preserved eggs are delicious, there are a lot of health risks that come with eating them. Therefore,

美食城堡的安全保卫战
A War to Protect Food Safety

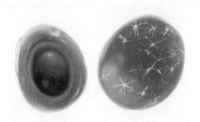

you should pace yourself and eat less of it, this is especially important to know for children, so please pay attention.

图 15　坏蛋皮蛋

Figure 15　Preserved Egg is a Bad Egg

美食城堡的安全保卫战
A War to Protect Food Safety

烧烤架上的大坏蛋
The "Bad" on the Grill

在盛夏的夜晚，美食城堡有一道"靓丽"的风景，那就是露天的烧烤店。狐大哥和他的一帮朋友们更是这里的常客，大家围坐在一起，吃着烤串、喝着啤酒好惬意。

这日，狐大哥又和黑野猪来吃烤串，突然发现烤串店都关门了。冲动的黑野猪打开手机，拨通了烧烤店的电话和老板理论却被告知烧烤店整修关门。次日，城堡内发出公示说这周日城堡将进行烧烤知识讲座。因为没有吃到烧烤而满肚子怨气的狐大哥义愤填膺地说："一

During summer nights, a "beautiful" scene will rise at night in Castle City, it is the outside barbeque skewer restaurants! Mr. Fox and his group of friends visit these skewer restaurants quite often. They all circle around the table and feast on drinks and barbecue, always having a wonderful time together.

One day, Mr. Fox and Mr. Boar came to have some skewers as usual, but they soon realized that all these stores were closed. Mr. Boar immediately called the BBQ restaurant owner on his cellphone to ask about the closed store. The boss told them that the restaurant was closed and under reconstruction. The next day,

117

A War to Protect Food Safety

定要参加讲座理论一下。"与此同时，小微博士受邀进行烧烤知识讲座。为了充分了解烧烤，课堂上小微博士就大家眼中的烧烤进行了调查，发现不少人喜欢烧烤，而且认为烧烤既然不会引发疾病那就是健康的。小微博士摇摇头表示否定，并告诉大家她这周日会在城堡的会议讲厅为大家揭晓答案，当然不忘提醒大家提前做功课。

周日很快就到了，在城堡卫生组织的大力宣传下吸引来了城堡的大部分居民。讲座开始，小微博士先拿出了一份检测亚硝酸盐的试纸

Castle City posted a notification about a "BBQ Safety Lecture", Mr. Fox was rather upset about not being able to eat BBQ, so he decided to go to the lecture. Mr. Micro was also invited to speak at the lecture. In order to fully understand BBQ, Dr. Micro conducted a small survey on everyone's opinion about BBQ. She realized that many people enjoy it, and most people thought that BBQ was a safe type of food that did not cause illnesses. Dr. Micro shook her head and told everyone that she would present an answer during the lecture on Sunday, and she also reminded everyone to do some homework and research beforehand.

It was soon Sunday, and the lecture has attracted majority of the city's citizens' attention and attendance. The lecture began with Dr. Micro holding

美食城堡的安全保卫战
A War to Protect Food Safety

条并向大家解释它的作用，然后将试纸条放入一串处理过的烤肉里，结果检测到了高浓度的亚硝酸盐。这时候大家炸开了锅，你一言我一语，狐大哥拍拍黑熊的肩膀道："老黑熊，亚硝酸盐是不是前段日子报道的那个致人中毒的东西呀？"黑熊讷讷道："好像是听到过这么一个事，听说这东西还能致癌。"这时小微博士示意大家静一静，并语重心长地告诉大家烤肉串不新鲜，大多有很高的亚硝酸盐产生，此外，肉串中有亚硝酸盐只是烤肉中一个小小的危害。小微博士紧接着放出了"米猪肉"的照片（一种感染寄生虫的肉），讲台前的雨燕突然道："我知道这种肉，这种肉是动物感染了寄生虫引起的，如果肉没有烧熟吃到身体里危害很大，很容易患病。"小微博士点点头表示

a nitrite test strip. She explained to everyone its usage and put the test trip onto a meat skewer, and the test result showed high concentration of nitrite. Everyone was blown away and began discussing about the test, Mr. Fox patted Mr. Black Bear on the back and said, "Hey buddy, wasn't there a report on nitrite as a type of salt that is poisonous?" Mr. Black Bear replied, "I recall hearing something like that, and apparently, this thing can also cause cancer." Dr. Micro then told everyone to settle down and began to explain to the citizens that if the BBQ skewers were not fresh, then they usually had a high concentration of nitrite, and even then, nitrite only posed as a small danger. Dr. Micro then showed a picture of "rice pork meat" (meat with

美食城堡的安全保卫战
A War to Protect Food Safety

赞同。另外小微博士还告诉大家烧烤食物偏向燥热，再加上孜然、胡椒、辣椒等都是辛辣刺激的食物，会大大刺激胃肠道蠕动及消化液的分泌，有可能损伤消化道黏膜，还会影响体质的平衡，令人"上火"。其实不仅如此，烧烤食物降低了营养物质的利用率；食用过多烧煮、熏烤的蛋白质类食物，如烤羊肉串、烤鱼串等，将严重影响青少年的视力，导致眼睛近视。

到了最后提问环节黑熊提问到，特别喜欢吃烧烤食物怎么办？小微博士回答道，烧烤食物并不是

parasites), Emily suddenly said, "I know this type of meat! This is the type of meat is from animals with parasites. If the meat is not cooked thoroughly, then it can easily make you sick." Dr. Micro nodded in agreement with Emily she also added that BBQ skewers are rather dry and all kinds of spices are added to it, therefore the skewers will affect our stomach to secret more digestive fluids that may harm our gastrointestinal mucosa. On the other hand, BBQ also lowers the nutritional values of the food. Protein foods that are over cooked or barbequed, such as lamb and fish-ball skewers, can harm one's eyesight and cause nearsightedness.

It was Q&A time and Mr. Black Bear asked what he should do if he just loved BBQ. Dr. Micro explained that he

A War to Protect Food Safety

不能吃，而是要适量吃、合理吃，并给出了以下建议：

1. 烧焦的部位不要吃。越烤黑的部位致癌物含量越高；

2. 多吃新鲜的蔬菜与水果。蔬菜水果中的抗氧化物质、植物化学物等可抑制有毒有害物质的致癌、致畸作用，对于降低其危害有积极的作用。例如猕猴桃、西红柿、橙子、胡萝卜、甘蓝、西兰花、菠菜等，既可以丰富餐桌、平衡膳食，还可以保护自己、降低危害；

3. 搭配膳食纤维含量丰富的食物一起吃。膳食纤维可吸附有毒有害物质，并将其带入粪便排出，防止有害物质被吸收进入血液。如玉

were not restricted from eating BBQ food, but he had to be careful not to eat too much, please eat a reasonable amount. She also gave some suggestions:

1. Do not eat the overcooked food, which may contain carcinogens.

2. Eat plenty of vegetables and fruits because they contain antioxidants and phytochemicals, which can help resist harmful substances that are poisonous or cancerous. Fruits and vegetables such as kiwi fruit, tomato, orange, carrot, broccoli, cauliflower, spinach and etc., they are all great to add to the table, they're good for your health and can also help protect your body.

3. Dietary fiber should also be added to your daily diet. Dietary fiber can help absorb toxic and harmful substances, and it also prevents harmful

A War to Protect Food Safety

米等粗粮、蔬菜、水果、菌藻等含膳食纤维多；

4. 能去掉外皮的，最好去掉外皮吃里面的肉，如烤鸡肉、烤鸭肉、烤鱼肉等；

5. 科学选择烧烤种类，减少热量和脂肪摄入。关于烧烤的肉类选择，推荐无骨鸡胸肉、鸭胸肉和瘦牛肉。尽管较多的油脂在烧烤过程中会产生诱人的香气，但还是建议少选择带皮或肥肉过多部位的肉，有些还可以去皮烤。比如鸡翅，去皮烤一样可以口味很鲜嫩，热量和饱和脂肪却大大减少；

substances from being absorbed into the bloodstream. Corn, oatmeal, vegetables, fruits, algae and etc. foods contain plenty of dietary fiber.

4. It is also best to eat the meat without the skin, if it's possible, always skin the meat beforehand, such as chicken, duck and fish meat.

5. Use the scientific way to select the right type of BBQ to eat, try to decrease your calorie and fat intake. You may select meats such as boneless chicken breast, duck breast or lean beef. Even though the fat within the meats will create a delicious aroma when it's being cooked on the BBQ, it is best to eat piece with less fat. You may also take the skin off and eat the meat. For example, you can skin the chicken wing, it is still delicious and it contains much less fat.

美食城堡的安全保卫战
A War to Protect Food Safety

6. 烤肉刷蒜汁和番茄酱可减少致癌物。如果是自制烤肉，还可以提前用蒜汁调味，也可以同时刷上番茄酱；

7. 烧烤时离火远一点，可减少致癌物产生；

8. 多吃些凉性食物和生蔬菜，可以去火并促进致癌物排出。

讲座结束，大家纷纷为城堡卫生组织的这次活动点赞，并提议多多举办类似的科普讲座。

6. Brushing garlic sauce or tomato sauce on the BBQ can also help decrease the amount of carcinogen in the food. If you are preparing BBQ yourself, you can brush some garlic sauce or tomato sauce beforehand.

7. Keeping the food away from the fire while cooking can also help decrease the amount of carcinogen.

8. Eating cold foods or raw vegetables on the side can also help discharge carcinogen.

The lecture ended, and Castle City's citizens' all applauded this event and thanked the efforts of organizing the event. They also hoped that there could be more informative lectures in the future.

A War to Protect Food Safety

美食城堡的安全保卫战

图 16 烧烤架上的食物

Figure 16　The food on the Grill

【小微博士有话说】

1. 提倡蒸、煮、炖、焯的食品加工方式，提倡绿色食品、清淡饮食及科学的加工方法。

2. 各种食物都不提倡生吃，尤其是肉食类食物。

[Dr. Micro's notes and tips]

1. It is best to steam, boil, slow-cook, or scald your food. Eating green and healthy is also encouraged.

2. Eating raw food is discouraged because it may contain harmful substances, especially raw meat.

美食城堡的安全保卫战
A War to Protect Food Safety

生的海产品，我们不约
Say No to Raw Seafood

世间唯美食与爱不可辜负。夏天到了，又到了油焖大虾、糖醋带鱼、清蒸螃蟹等闪亮登场的时候了，吃货们也蠢蠢欲动了。

周末到了，早晨胖胖熊跟着妈妈去美食城堡里的超市购物。一进超市，胖胖熊就开始东张西望，不一会儿就跟熊妈妈走散了，熊妈妈专心购物也无心管他。一个多小时过去了，熊妈妈把该买的东西都挑好后开始去找胖胖熊。她绕了超市好大一圈，终于在生鲜区找到了胖胖熊，胖胖熊正盯着生鲜柜里的海产品，看得可认真了，以至于都没

Summer time has arrived, and it is the season for fried shrimp, sweet and sour fish, steamed crab and other magnificent seafood dishes. The foodies in town were eager to eat.

It was a morning on a weekend, and Little Bear went to the supermarket with Mother Bear. As soon as they arrived, Little Bear started wondering around and he was separated from Mother Bear. Mother was too busy with her shopping to notice either. After an hour, Mother Bear has bought everything she needed so she began to look for Little Bear. She walked around the supermarket,

美食城堡的安全保卫战
A War to Protect Food Safety

听到身后不远处妈妈的呼唤声，直到熊妈妈走近用手指敲了一下他的脑袋瓜，他才回头。一看到是熊妈妈，胖胖熊立刻兴奋起来，嚷着要熊妈妈买些海产品回家。熊妈妈为难了，因为她不会烧海鲜呀，而且熊爸爸也不会，买回去没人做呀。熊妈妈决定不买，可是胖胖熊坚持要买，嚷着不买就不回家。在胖胖熊的强烈要求下，熊妈妈最终还是买了，她也想买回家学着做。他们买了螃蟹、大虾还有甲鱼，熊妈妈本不想第一次就买这么多，可是胖胖熊一定要挑这么多，熊妈妈觉得自己压力可大了，因为要一下子学这么多种海鲜的做法，还真是一件难事。

回家的路上，胖胖熊可开心

and finally found Little Bear around the seafood isle, he was staring at the seafood in the freezer. Little Bear turned around and saw Mother Bear, he was very happy and urged her to buy some seafood home. Mother Bear was troubled because she didn't know how to cook seafood, and neither did Papa Bear. But Little Bear was rather persistent, so Mother Bear finally bought some seafood home because Little Bear said he would not go home until they buy seafood. They bought crab, prawns and a soft-shelled turtle. Mother Bear didn't intend to buy so much, but Little Bear really wanted it. Mother Bear was under a lot of pressure, how could she learn to make all these seafood at once? It was an almost impossible task.

Little Bear was very happy on their

美食城堡的安全保卫战
A War to Protect Food Safety

了，想着今天家里要烧海鲜，他的口水都快止不住了。到家后，熊妈妈就开始在厨房里忙活起来，胖胖熊就在一边玩。玩着玩着胖胖熊突然想起松仔来，自从上次小微博士教育他要学会分享和感恩后，胖胖熊一直都记着，松仔平时帮了他不少忙，今天他想邀请松仔来家里吃饭，和松仔一起品尝妈妈做的海鲜。于是他跑到厨房告诉妈妈，今天他要邀请松仔来家里吃饭，熊妈妈听到后先是有点诧异，然后就笑了，夸胖胖熊长大了懂事了。胖胖熊有些害羞地说："那我去找松仔了，妈妈你加油，我们回来就开饭哦，"然后扭头跑了。熊妈妈微笑着继续忙手里的活。

way home, because he would be having seafood, just the thought of it made him drool. Once they arrived home, Mother Bear started to cook in the kitchen, and Little Bear began to play on the side. Suddenly, Little Bear thought of his friend Squirrel. Dr. Micro has been teaching them about being thankful and sharing with friends. Little Bear remembered that Squirrel had been helping him a lot, so he thought he should invite Squirrel over for dinner to share the seafood his mom cooked. Little Bear ran to the kitchen to tell his mother that Squirrel was coming over to eat. Mother Bear was initially surprised, then she smiled, for Little Bear had grown up. Little Bear bashfully said: "I'm going to invite Squirrel then, thank you mom." Then he turned and

美食城堡的安全保卫战
A War to Protect Food Safety

　　松仔在胖胖熊的邀请之下，来到胖胖熊家。熊妈妈和熊爸爸很是欢迎松仔，熊妈妈知道松仔很聪明，这不又开始想向松仔讨教了。熊妈妈正在厨房里清蒸螃蟹，但她自己又看不出来到底有没有熟，于是问松仔有没有经验。松仔看了一下也不是很确定，并解释自己不是很懂烹饪，但是之前在美食城堡参加过一次食品安全宣讲会，说是海产品一定要煮熟，不然不能吃。熊妈妈一听被吓倒了，她不确定自己刚做的其他海鲜有没有煮熟，觉得这下子麻烦了，不敢吃，丢掉又很浪费，而且今天午饭就没下饭菜了，还好松仔聪明，告诉熊妈妈其实还有补救的方法，那就是把所有菜都再适当煮一下，确保完全熟了

left. Mother Bear kept on working with a smile on her face.

　　Squirrel came to Little Bear's house upon his invitation. Mother Bear and Papa Bear was fond of Squirrel because Mother Bear knew that Squirrel was smart, and she had lots to ask him. Mother bear was steaming crabs in the kitchen, but she couldn't tell if they were ready or not, so she asked Squirrel for help. Squirrel took a look, but he wasn't sure either, and he explained that he was not familiar with the culinary arts, but he once attended a seminar on food safety in Castle City, and he knew that seafood must be cooked completely, or else they couldn't be eaten. Mother Bear was distraught upon hearing this, as she couldn't be sure whether she had completely cooked

A War to Protect Food Safety

再起锅。于是熊妈妈按松仔的说法做了，大家才放心开饭。饭后松仔说自己有很多疑问，他想去请教一下小微博士，胖胖熊也很感兴趣，于是他俩一起前往小微博士家。

小微博士正在给自家的花浇水，老远就听到胖胖熊在叫自己。

the food before. Now this was a messy situation, on one hand they're not sure whether the seafood is ready to eat, and on the other hand, it's rather wasteful to throw it all away. Mother Bear had only prepared seafood for the night. But Squirrel was smart, so he assured Mother Bear that there were still ways to salvage the meal, they could cook all the dishes again a little more to make sure everything was thoroughly cooked. So, Mother Bear followed Squirrels suggestion, and everyone ate the dishes with a peace in mind. After the dinner, Squirrel had a lot of questions in mind, so he wanted to go find Dr. Micro. Little Bear was interested too, so the two friends went to Dr. Micro's house together.

Dr. Micro was watering her own plants, and she heard Little Bear calling

美食城堡的安全保卫战
A War to Protect Food Safety

小微博士请胖胖熊和松仔到家里喝茶，松仔开始问问题："博士，为什么海鲜必须保证完全煮熟后才能吃？"小微博士开始详细地解释："第一点：现在大海污染都比较严重，所以打捞的海产品中有寄生虫，如果没有经过高温煮熟，无法彻底杀死，人食用后会感染寄生虫；第二点：海产品很难保证不受到细菌污染，进食了被副溶血性弧菌污染的海产品是很危险的，轻者腹痛腹泻，重者休克昏迷甚至失去生命，而大多数病菌在高温下很容易死亡；第三点：生的海产品中含有一种硫胺素酶，会分解破坏食物中的维生素B1，如果加热到60摄氏度以上时，硫胺素酶就会失去作用；第四点：生的特别是新鲜的海产品中含有组氨酸，易引起人体的过敏反应；第五点：一些软体贝类

her name from far. Dr. Micro invited Little Bear and Squirrel inside her home for tea, and Squirrel began asking questions immediately. "Dr. Micro, why is it that seafood must be completely cooked before eating?" Dr. Micro began explaining the reasons in detail, "First of all, the ocean is heavily polluted right now, so a lot of the seafood caught in the sea contains parasites. If it is not boiled in high temperature thoroughly, then the parasites cannot be killed, and you will be infected as well if you eat it. Secondly, it is difficult to ensure that seafood is not contaminated by bacteria, eating seafood contaminated by Vibrio parahaemolyticus is very dangerous to one's health, it can cause abdominal pain and diarrhea, and even severe shock and coma that can lead

A War to Protect Food Safety

中含有毒素，对人体有很大的毒副作用，会出现腹泻，甚至会刺激到人类的神经系统。

好了，以上是解释了生的海产品为什么不能吃，然后你们既然来了，并提到了海产品，那我就再给你们普及一些知识吧。生的水生植物也不宜吃，如菱角、荸荠。

to death. However, most bacteria can be eliminated under high temperature. Thirdly, raw seafood contains a type of thiaminase that can destroy the Vitamin B1 content within foods, but if you heat the seafood up to 60 degrees Celsius, then the thiaminase will be eliminated. The fourth point is that fresh seafood usually contains histidine, and it can easily cause an allergic reaction. The fifth and final point is that some soft shellfish contains toxins as well, and it can cause rather harmful side effects to the human body, such as diarrhea, and it can even effect the human nervous system.

Alright, now all the explanations that I just gave are reasons to why you cannot eat raw seafood. But since you are here already, let me tell you some general knowledge. Raw sea vegetation

A War to Protect Food Safety

藕、茭白等，水生植物容易被细菌污染，过多食用生的或者没有煮熟的菱角，极易引起肠道等消化系统等疾病。死的海产品也不宜吃，如死虾、死螃蟹等，因为死掉的大多数海产品会产生组胺，组胺是一种有毒物质，并且随着死亡的时间延长，海鲜体内的组胺积累越来越多，毒性越来越大，即使煮熟了，这种毒素也不易被破坏。最后，最好不要烹食新鲜的河豚，河豚肉质虽然鲜嫩肥美，但其皮、内脏、眼、血液等含大量毒素，尤其在产卵期毒性就更大，其毒素不能被盐腌、日晒、烧煮等破坏，食后多数会使人中毒。好了，今天就给你们讲到这了，讲太多，怕你们记不住。"胖胖熊都快睡着了，不过他虽然没记住啥啥不能吃的原因，但是他记住了哪些不能吃，于是和松

should not be eaten, such as water chestnuts, lotus roots, zizania aquatic and etc. Aquatic plants can easily be contaminated by bacteria, too much raw or uncooked water chestnut can easily lead to intestinal and other digestive diseases. Dead seafood should be paid extra attention as well, such as dead shrimp, dead crabs and others. Most dead seafood will produce a toxic substance called histamine, and dead seafood will accumulate more this toxic substance overtime. The amount of toxins can accumulate, where the toxic substance within the seafood gets more amd more, to the point where cooking the seafood does not help release the toxicity. Finally, it is best not to cook puffer fish, even though fresh puffer fish is delicious, its skin, intestines, eyes and

仔谢过小微博士后就开心地回家了。

blood contains a large amount of toxins, especially during its spawning period. The toxins cannot be eliminated through salting, exposure under the sun, nor does cooking it take away its toxins. Okay, that is all I will say today, you won't remember everything if I go on with more facts." Little Bear was falling asleep, even though he did not remember the reason behind why he could not eat certain foods, he at least remembered which items to stay away from. Squirrel and Little Bear thanked Dr. Micro for the information, and went home happily.

图 17　生的海鲜

Figure 17　Live seafood

美食城堡的安全保卫战
A War to Protect Food Safety

【小微博士有话说】

1. 生的海产品不宜吃，原因总结起来有四点：第一，长期生吃海产品会缺乏维生素B1；第二，有感染寄生虫的危险；第三，有感染致病菌的危险；含有毒素易引起中毒；第四，易引起过敏。

2. 生的水植物不宜吃，容易引起肠道等消化系统疾病。

3. 死的海产品不宜吃，死的海产品产生的组胺有毒。

4. 河豚有毒且毒性大，最好不要吃。

[Dr. Micro's notes and tips]

1. It is best not to eat raw seafood for four main reasons: First, eating raw seafood long term will cause Vitamin B1 deficiency; Second, there is danger of being infected by parasites; Third, there is danger for bacterial infection or being poisoned; Fourth, it can cause allergies.

2. Raw Sea vegetation should not be eaten because it can cause diseases in the digestive system.

3. Dead seafood should not be eaten, because dead seafood can produce poisonous histamines.

4. Puffer fish is highly poisonous, it is best not to eat it.

美食城堡的安全保卫战
A War to Protect Food Safety

香蕉宝宝的爱与恨
Love and Hate of Banana Baby

在美丽的美食城堡中，每到特定的时节都会有大量的时令蔬菜水果上市，不仅新鲜而且价格还特别便宜。这不又到了香蕉大量上市的时候，熊妈妈看一家人都比较爱吃香蕉，而且价格十分的便宜就买了一箱香蕉回家，为了可以让香蕉保存更长的时间，熊妈妈只挑了一部分黄皮的香蕉以便这两天吃，剩下的大部分都是特地挑选那些皮发青的香蕉。

熊妈妈开开心心地带着一大箱的香蕉回到了家，推开门看到胖胖熊、松仔和雨燕三个好朋友都在家里面玩耍，就跟他们几个说："胖

In the beautiful Castle City, there are plenty of fresh and cheap seasonal fruits and vegetables sold in the markets. It was the season when bananas were ripe. Mother Bear went to the market and decided to buy a box of bananas, since they were cheap and her family enjoyed bananas. The box of bananas she bought was not only filled with yellow-skinned banana, but also green-skinned bananas, because the green-skin bananas could stay fresh longer.

Mother Bear brought the box full of bananas back home with joy. She went home and found Little Bear, his friends Squirrel and Emily playing together.

美食城堡的安全保卫战
A War to Protect Food Safety

胖熊你们几个先别玩了，先吃点香蕉吧！"胖胖熊就屁颠屁颠地跑过来，把妈妈买回来的黄皮香蕉拿去和松仔、雨燕两个好朋友分享。胖胖熊一边抱着香蕉皮一边说："香蕉是我最喜欢吃的水果啦，好开心啊！"松仔也一边咬着胖胖熊递过来的香蕉一边说："香蕉不仅香香甜甜很好吃，对我们还有很多好处呢。比如预防高血压，消除疲劳，预防便秘，防治胃溃疡，防治失眠等。"雨燕崇拜地看着松仔说："松仔，你好厉害呀！懂得香蕉的这么多好处。"胖胖熊也连连点头："对呀对呀，松仔你好厉害，既然香蕉有这么好处，我可要多吃点。"

过了几天，胖胖熊发现妈妈买回来的黄皮香蕉已经被吃完了，还

Mother Bear said, "Little Bear, would you like to eat a banana?" Little Bear happily accepted the offer and shared the bananas with his two friends. "Banana is my favorite fruit!" exclaimed Little Bear as he quickly devoured his banana. "Bananas are not only tasty, but they also have a lot of health benefits," said Squirrel. "For example, banana can counter high blood pressure, alleviate stress and fatigue, prevent constipation, stomach ulcer, insomnia, and many more benefits." Emily looked at Squirrel with amazement and said, "Wow, you know so much about bananas, Squirrel!" Little Bear chimed in "yeah that's right, you know so much, Squirrel! I'm going to eat more bananas from now on."

Few days later, Little Bear had eaten all of the yellow-skinned bananas,

美食城堡的安全保卫战
A War to Protect Food Safety

有好多是半青不黄的香蕉,捏一下还有点硬,不像之前吃的香蕉软软的,不过转念一想,先尝尝,看看味道怎么样。于是胖胖熊就拿了一根香蕉尝了一下,除了稍微有一点点涩和硬其他没什么差别。胖胖熊就开始吃这些半青半黄的香蕉了,而且每天还吃好多根。一段时间后,胖胖熊觉得最近一段时间老是感觉便秘,肚子也很不舒服。松仔得知情况后感觉很奇怪,觉得胖胖熊不应该这样,于是就一起去请教小微博士。博士听了两人的叙述,并询问了胖胖熊的饮食习惯后,就说:"松仔你说的都很好,吃香蕉是有很多好处,可以预防便秘,但是生香蕉的涩味来自于香蕉中含有大量的鞣酸。当香蕉成熟之后,虽然已尝不出涩味了,但鞣酸的成分仍然存在。鞣酸具有非常强的收敛

and only the green-skinned bananas were left. Although the green-skinned bananas felt a little bit harder, and they tasted a little bit unsmooth. Little Bear decided to go ahead and eat the green-skinned bananas too. After a few days, and eating a lot of bananas, Little Bear felt constipated and his stomach did not feel good at all. When Squirrel found out about Little Bear's situation, he was confused as well. Therefore, they went to Dr. Micro to find the answer. After she listened to their story, Dr. Micro said, "What Squirrel said about bananas are true, eating bananas have a lot of benefits. But the unsmooth-like tastes of the green-skinned bananas are from tannic acid in the banana. You can't taste the tannic acid in ripped and yellow banana, but it is still there.

美食城堡的安全保卫战
A War to Protect Food Safety

作用，可以将粪便结成干硬的粪便，从而造成便秘。最典型的是老人、孩子吃过香蕉之后，非但不能帮助通便，反而可能发生明显的便秘。此外，多吃香蕉还会因胃酸分泌大大减少而引起胃肠功能紊乱和情绪波动过大。因此，香蕉不宜过量食用。而且也不可空腹食过多的香蕉，因为香蕉中含有大量的钾、磷、镁，对于正常的人，大量摄入钾和镁可使体内的钠、钙失去平衡，对健康不利。"松仔和胖胖熊听了小微博士这番话，决心以后就是再好吃再有益的食物也不能多吃。

Tannic acid can harden feces, making people feel constipated. This usually occurs when elders and children consumes banana. Eating a lot of bananas can also limit gastric acid secretion, which can disrupt the digestive system. That is why you shouldn't eat too much bananas. This is especially the case when you eat banana on an empty stomach, because banana contains a lot of potassium, phosphorous, and magnesium, and a lot of chemical consumption of those chemicals can disruption the sodium and calcium in the body." After listening to Dr. Micro, Squirrel and Little Bear decided that it was best not to eat too much of any one food item, even if it was good for your health.

美食城堡的安全保卫战
A War to Protect Food Safety

图 18 香蕉熟了吗

Figure 18　Is the banana matured?

【小微博士有话说】

1. 香蕉的好处多，预防高血压，消除疲劳，预防便秘，防治胃溃疡，防治失眠等样样都可以。

2. 香蕉的禁忌是不宜多吃，不宜空腹吃，不宜生吃。

3. 适量吃香蕉，健康伴你行。

[Dr. Micro's notes and tips]

1. Bananas have many health benefits, including preventing high blood pressure, constipation, stomach ulcer, sleeplessness and relief stress.

2. Do not eat too many bananas, and never eat it when it is unripe or when on an empty stomach.

3. If you eat an appropriate number of bananas, then it will help you stay healthy.

·139

美食城堡的安全保卫战
A War to Protect Food Safety

香甜诱惑

Temptation of Sweet

中午知了不知疲惫地叫着，给人带来一种夏日的烦躁，没有一丝风，大地活像一个蒸笼。美食城堡的松仔、胖胖熊和雨燕三个小伙伴结伴走在回家的路上，被这炙热的空气烤得无精打采的。走着走着，他们路过了一家甜品店，看到店家挂在外面海报上展示的冰激凌，想象着冰激凌进入嘴巴冰冰凉凉、香甜可口的画面。于是胖胖熊按捺不住，率先跑到甜品店的窗口准备买冰激凌，松仔和雨燕看到胖胖熊这一举动，也紧跟着进入甜品店的橱窗前准备买冰激凌。不一会儿，几个人就一人拿着一只冰激凌吃着，感觉走路都有精神了。胖胖熊边吃

It was a very hot day in Castle City. Squirrel, Little Bear, and Emily felt very hot under the blazing sun. The three friends passed by a sweet shop that sold ice cream cones, they were very tempted to have some, especially in such a hot day. Little Bear ran up to the shop to buy ice cream, Squirrel and Emily soon followed. After eating the ice cream, the three friends felt much cooler. As they were eating, Little Bear said, "Ice cream is delicious, I wish I have more allowance so I can buy more." Squirrel said, "Ice cream is great in hot weather, but I ran out of my allowance too. We should go home and ask our

美食城堡的安全保卫战
A War to Protect Food Safety

边说："冰激凌可真好吃，这么热的天气真想多吃几根，可惜没有零用钱了。"松仔也说："这冰激凌冰冰凉凉的，又解渴又解暑，吃了浑身都有力气。就是没有钱了，不然我也想多买几根。要不我们回去和爸爸妈妈商量一下，现在夏天天气这么热，每天中午都要走这么远的路上下学，给我们加点零用钱好买解渴的冰棒和冰激凌。"胖胖熊和雨燕听了之后都表示同意。

于是，三人回到各自的家中和父母商量增加零用钱的事情。各家的父母考虑到最近天气确实很热，让他们自己多吃点冷饮也不错，就同意了。接下来这一个月的炎炎夏日，松仔、胖胖熊和雨燕三个小伙伴在每天中午放学时都会买两三只冰激凌在路上吃。渐渐地，胖胖熊的爸爸妈妈发现胖胖熊变得更胖

parents for more allowance to buy ice cream." Little Bear and Emily agreed with Squirrel.

Therefore, the three friends went back home and asked their parents for more allowance. Considering that the weather was so hot, their parent agreed to increase their allowance. For the whole month during the summer, Squirrel, Little Bear, and Emily would always buy a few servings of ice cream and eat it when they walked back home from

141

美食城堡的安全保卫战
A War to Protect Food Safety

了，松仔和雨燕的爸爸妈妈觉得自己的孩子越来越瘦，而且每次回家吃饭都吃一点点。几位爸爸妈妈聚在一起商量孩子们到底是因为什么事情变成这样的。熊爸爸说："我们去找小微博士吧，她一定会知道为什么。"

小微博士听完这些爸爸妈妈们的问题，又询问完最近一个月孩子们的饮食状况后说："胖胖熊变胖，松仔和雨燕变瘦都是由于过量食用冰激凌。冰激凌的热量很大，胖胖熊是那种消化功能好的，容易吸收，再加上正常饮食，体重自然而然就会增加。但是像松仔和雨燕这样的，大量食用冰激凌后，会使胃肠功能紊乱，厌食毛病会越来越重，影响营养吸收，体重下降。"爸爸妈妈们听完小微博士的话后，才知道都是平时吃多了冰激凌的原

school. After some time, Little Bear's parents noticed that Little Bear became plumper, while Squirrel's and Emily's parents noticed that their child became thinner. The three groups of parents were confused, and decided to all go and ask Dr. Micro.

"This is because of the ice cream." explained Dr. Micro, after she listened to their story, "Little Bear is good at absorbing nutrient, so the extra ice cream made him plumper, while Squirrel's and Emily's digestive system are disrupted by the ice cream and it causes them to lose their appetite, eat less, and eventually lose weight." Dr. Micro continued to explain, "It can get very hot during summer, and the heat causes people to lose their appetite, eating some ice cream can help, but

A War to Protect Food Safety

因。小微博士接着说："在夏天的时候天气十分的炎热，吃饭时有时候会没有胃口，吃一些冰激凌，是一个迅速补充体力降低体温的好方法。但是要控制量，每天吃一到两只即可。"

爸爸妈妈回去后就对自己的孩子进行了批评教育，规定以后每天只能吃一个冰激凌，并且到了饭点必须吃饭。

don't over eat."

After they found out the reason behind the weight change in their children, the parents went back home and limited the amount of ice cream allowed each day, and also instructed the children to eat regular meals on time.

图 19　冰激凌

Figure 19　Ice cream

美食城堡的安全保卫战

A War to Protect Food Safety

【小微博士有话说】

1. 夏天吃冰激凌是一个迅速补充体力降低体温的好方法，但是要适量。

2. 过量食用冰激凌会导致肠胃伤害、发胖甚至会厌食，因此不要因为贪吃而伤害自己的健康。

[Dr. Micro's notes and tips]

1. Eating ice cream in the summer is a great way to replenish energy and stay cool, but the amount consumed should be controlled.

2. Too much ice cream can disrupt the digestive system and make you gain weight or loose appetite.

美食城堡的安全保卫战
A War to Protect Food Safety

虫眼＋果蔬＝绿色？
Bug holes+Vegetables=Green?

近日，雨燕发现美食城堡的居民们画风大变——对带虫眼的果蔬情有独钟。在清晨的农贸市场上，不难看见人们在各摊位上搜寻着这种果蔬，如同寻宝。而在以往，人们在购买水果、蔬菜时都喜欢挑个儿大、颜色鲜艳、表皮完好的。这是为什么呢？原来，随着生活品质的不断提高，人们对食物的追求不再停留于是否能够填饱肚子的问题上，而是更加关注其营养价值、安全性。

但是，果蔬的农药残留一直让人们耿耿于怀。不知何时起，"果蔬带虫眼，说明有虫；虫子能

Lately, Emily noticed a change in Castle Town, everyone was buying vegetables and fruits with bug holes in them. You can see people looking for fruit items with holes each early morning in the vegetables and fruit market. Before, people used to buy fruits and vegetables that were big, well-proportioned and with beautiful colors. So why was this happening? It is because that people are looking for foods that not only can fill their stomachs up, but also "safe" and reliable.

There has been a rumor around saying that bugs will only eat fruits and vegetables that are not treated with

A War to Protect Food Safety

存活,说明没使用农药;不喷洒农药,说明果蔬就更绿色安全"的说法在人们口中传了开来;再加一些个体商贩的借机炒作,"我们的青菜本来是给自己吃的,没有打过农药……"。因此,带有虫眼的果蔬便成了人们判定"绿色果蔬"的"金标准"。当然,这些所谓的"绿色果蔬"价格也自然会高些,但在市场上仍是供不应求。

那么问题来了,虫眼果蔬就真的是绿色安全的吗?

雨燕为解开这个疑惑,走遍大大小小的农贸市场,终于淘到了些虫眼蔬菜。与此同时,她还随机买了些表面完好的蔬菜,一同带去给小微博士检测,但结果却令她瞠目结舌。这些虫眼蔬菜上的农药残留量远远高于表面完好的蔬菜。相比雨燕一脸的诧异,小微博士显然对

pesticide. Therefore, people began to look for the signs of bug bites on fruits and vegetables, they regard the signs as proof of a "green" and healthy food item. Of course, such "green" items are also priced higher than usual, and it is in demand as well.

But is it really true?

Emily wanted find out the truth about this rumor. She walked around the market, and bought some items that had holes from bug bites. She also bought normal fruits and vegetables that had no holes. She brought all of them to Dr. Micro to test out the difference. The result surprised Emily; the items with

A War to Protect Food Safety

这结果并不感到意外，因为在早些时候，她就亲自走访过当地的果蔬种植大户，并对他们的果蔬管理模式有了一定的了解。

在农业生产过程中，病虫害预警讲究"防患于未然"。以青菜为例，菜叶上有虫眼说明病虫为成虫，而那些没有虫眼的青菜可能存在幼虫，但成虫的抗药能力一般比幼虫要强，那么在使用农药过程中，虫眼蔬菜的农药含量肯定会高一些；其次，病虫长大需要一个过程，等青菜出现虫眼后再使用农药，这间隔距蔬菜采收、售卖时间往往过短，大部分农药还没有来得及分解，导致蔬菜上的农药残留量就高了；再次，喷洒农药很难做到

holes had way more pesticide remains than the normal fruits and vegetables. Contrary to Emily's surprise, Dr. Micro seemed to have predicted the result already, because she visited plantations and farms before and understood the harvesting process.

During the process of agricultural productions, the way to prevent pests is to "take preventive measures early on". Take vegetables for an example, the bites and holes are probably leftby mature pests, while the greens without any holes may contain pest larvae. Matured pests are general more resistant to pesticide content than larvae. Therefore, a larger amount of pesticides will be used on vegetables with holes than other. On the other side, pests normally go through a growth process in order to

147

美食城堡的安全保卫战
A War to Protect Food Safety

在同一时间杀死全部病虫，菜农为"抢救"遭虫害的青菜，通常会在短时间里加大剂量多次喷药，而农药自然降解的时间至少需要6~7天，这样的做法不但会使农药毒性叠加，久而久之还会使害虫产生耐药性，加大其危害性；最后，果蔬的表皮有一层蜡质，能有效地起到防止虫害和有毒物质的侵害，一旦其表皮有了虫眼，这层天然屏障就被破坏了，各种病原微生物就会乘虚而入，造成果蔬污染。

mature, therefore if pesticide is used when there are already bug bites on the vegetable leaf, then it might be too late. The time left for harvest and sales is often too short if the pesticides are being used when bug bites appear, therefore most of the pesticide residue on the vegetables have not time to break down. This results in pesticide residue left on vegetables. In addition, it is hard to kill all pests at once during the same time, so most farmers will spray the vegetables several times in a short period of time in attempt to save the vegetables. This will makes the pesticides even more toxic. Over time, the pest will build up resistance against pesticide, and the residue becomes increasingly harmful to humans. Finally, there is a special layer of wax on the

经小微博士这么一分析，雨燕恍然大悟。"果蔬上有虫眼，并不等于没使用农药，更不能作为挑选'绿色果蔬'的金标准。小微博士，那'绿色果蔬'通常如何辨别呢？"雨燕瞪大了眼睛问道。

"绿色果蔬首先会有一个'绿色'认证标签；并且在购买过程中，尽量选择当季果蔬，有的应季果蔬不仅营养价值高，而且农药使

skin of fruits and vegetables. This layer can effectively prevent pest bites and harmful substances from entering. If this layer of natural barrier is destroyed by pest bites, then all sorts of bacteria and other pathogenic microorganisms may take advantage of this, resulting in contaminated fruits or vegetables.

Through Dr. Micro's explanations, Emily finally understood it all. "The bug holes on fruits and vegetables are not proof of pesticide-free, neither can it be the golden standard to select your fruits and vegetables. Dr. Micro, what is the true way to identify 'green' food items?" Asked Emily.

"Green fruits and vegetables will have a 'green' verified label on them. It is also best to select fruits and vegetables that are in season. In

美食城堡的安全保卫战
A War to Protect Food Safety

用量也会相对低很多；形状、颜色奇怪的果蔬一般也不是'绿色'的，可能会有激素、化肥等有害成分；当然，多了解易受虫害的果蔬品种能更好地帮我们买到更安心的果蔬。"

"真是受益匪浅呢，我一定要把学到的告诉大家！"，还未等小微博士的话音落下，雨燕便激动地纵身一跃，扑打着翅膀就离开了。

season fruits and vegetables are very nutritious, and it will also contain less pesticides. Vegetables and fruits that are weird shaped and colored are not signs of 'green' either. They may contain hormones, chemical fertilizers or other harmful substances. Of course, understanding more about fruits and vegetables that are susceptible to bugs can also help you better select 'safe' food items."

"I've learned about so much today! I will let everyone else know of all the things I have learned about fruits, vegetables and pesticides today." Emily was so excited that she flapped her wings and immediately flew out to meet her friends.

A War to Protect Food Safety

图 20　有虫眼的果蔬

Figure 20　Vegetables and fruits with bug holes

【小微博上有话说】

在生活中，我们虽然对"绿色果蔬"的判定很难把握，但对于农药残留问题倒是可以持"宁可信其有，不可信其无"的保守态度。

那么，该如何减少果蔬表皮的农药残留量呢？下面，小微博士支4招，帮你轻松搞定！

[Dr. Micro's notes and tips]

It can be hard to determine what is "green and organic" in our daily lives, but with regards to pesticide residue, it's good to take an attitude of "guilty unless proven otherwise".

So, how can we reduce the amount of pesticide residues in our vegetable and fruits? Dr. Micro came up with four helpful tips to help you!

151

美食城堡的安全保卫战
A War to Protect Food Safety

1. 去皮。这是最简单也是最有效的方法，但通常只适合能去皮的瓜果类，如黄瓜、甜瓜等。

2. 水洗。不宜去皮的果蔬要充分浸泡、洗涤后食用；并且，不同果蔬应该采用不同的洗涤方法。例如叶菜类直接浸泡就能起到减少农药的效果，反复揉搓反而会破坏其表面蜡质的隔离作用；再如苹果、青椒等较厚实的果蔬，最好用流动水冲洗；有些果蔬还可采用碱水浸泡法清洗。

3. 加热。有些农药在高温时易挥发或分解，如氨基甲酸酯类杀虫剂，因此对于扁豆、菜花等蔬菜可

1. Peeling. This is the most simple and effective method, usually it's only suitable for fruits that can be peeled such as melon and cucumber.

2. Wash thoroughly. For fruits and vegetables that cannot be easily peeled, they should be washed thoroughly before being eaten; different fruits and vegetables should be washed in different ways as well. For example, soaking leafy greens in water is enough to reduce pesticides, but rubbing it might damage its surface wax. Apple, green pepper and other heavy fruits and vegetable must be rinsed thoroughly. Some fruits and vegetables can be immersed in salt water for cleansing.

3. Heating. Some pesticides can be evaporated or decomposed in high temperature, such as carbamate pesticide.

A War to Protect Food Safety

采用充分加热或事先水焯的方式，降低农药残留量。

4. 放置。研究表明，对于某些农药，暴晒一天即可使残留量下降一半。因此，像辣椒这种耐晒的蔬菜，完全可采用阳光直射的方法。

最后，小微博士建议，尽量少使用果蔬清洗剂、洗涤灵等清洁剂。虽然它们或多或少地能去除部分农残，但这些会对果蔬造成二次污染，因为它们本身所含有的一些物质，如表面活性剂等，对人体消化道会有一定的伤害。

So, for lentils, cauliflower and other vegetables, they can be heated prior to being eaten to reduce pesticide residue.

4. Sunlight. Studies have shown that certain pesticides can be reduced in half by simply putting them aside and allowing them to be exposed to sunlight. Therefore, vegetables like pepper that be can cleansed through putting it under the sun.

Lastly, Dr. Micro recommended everyone to use less detergent when cleaning fruits and vegetables. Even though they can help reduce pesticide residue, they can also cause a second contamination to fruits and vegetables, because they contain damaging substances such as surfactant. Surfactant is harmful to human digestive tract to a certain degree.

美食城堡的安全保卫战
A War to Protect Food Safety

炸鸡的正确打开方式
Eat Fried Chicken with Right Way

地上的土块被晒得滚烫，偶尔吹过的风中也充斥着黏稠感，天气闷热得很，盛夏就这样到来了。美食城堡的居民们，经历白天的燥热过后，最喜欢点上一份炸鸡，再配一扎冰啤酒，三五成群地坐着，肆意享受着难得的清凉。

胖胖熊是炸鸡的铁杆粉儿，哪怕是在饭后也总是能把一整份炸鸡消灭得干干净净。临近三伏天，胖胖熊的食欲愈发下降，但唯独对炸鸡的热爱只增不减，于是熊妈妈干脆给他点了两份炸鸡。一大瓶冰可乐，好让胖胖熊填饱肚子。这可把胖胖熊乐坏了，沁人心脾的冰可乐。香脆鲜嫩的炸鸡简直就是他心

Summer is here, when the rocks on the ground is boiling hot, and the air is humid and sticky. The citizens of Castle City love to order fried chicken and ice-cold beer at night, after a hot summer day.

Little Bear is fried chicken's biggest fan. No matter how much he has had for a meal, he's still able to eat an entire serving of fried chicken. It's been hot and humid recently, and even though Little Bear's appetite hasn't been great, his love for fried chicken hasn't abated. Mother Bear ordered two servings of fried chicken and a large

A War to Protect Food Safety

目中的完美搭档。一顿胡吃海喝之后，胖胖熊心满意足地去睡觉了。

可没等熊妈妈躺下，就听见胖胖熊突然喊肚子痛。只见胖胖熊蜷缩着身子，疼得在床上直打滚儿，额头上布满了细密的汗珠，吓得熊妈妈急忙把胖胖熊送到医院。经过一番检查后，发现原来是急性肠胃炎在作怪。医生解释道，饮食不卫生、冷热刺激或者暴饮暴食等，会损害胃肠黏膜，导致胃功能紊乱，极易造成急性肠胃炎，而在夏天则更应该注意饮食问题。熊妈妈连连

cup of coke so that Little Bear could fill his stomach on this hot summer day. Little Bear was ecstatic! A cup of ice cold coke, paired with delicious and crispy fried chicken, it was the ultimate deliciousness and the best meal he could ever wish for! After eating all the chicken and coke, he went to bed with satisfaction.

Right when Mother Bear was about to sleep, she heard Little Bear crying about his stomach hurting. Mother Bear was extremely worried, as she saw Little Bear crawled up in a ball on his bed, panting and sweating from his stomachache. They quickly went to the hospital, and after a series of check-ups, they found out that Little Bear had acute gastroenteritis. The doctor explained to them that eating

美食城堡的安全保卫战
A War to Protect Food Safety

点头，回家后赶忙向小微博士请教自己对胖胖熊的饮食管理方法。

小微博士在听熊妈妈道明整个经过后，可真是替胖胖熊捏了一把汗，好在胖胖熊现在并无大碍了。她语重心长地对熊妈妈说："在炎热的天气里，炸鸡火了，但我们的身体可不能'上火'呀。"

炸鸡属于典型的油炸食品，美

unsanitary foods or eating too much could damage the stomach's mucosa, disturbing the stomach's functions and ultimately causing stomachache and acute gastroenteritis. Especially in the summer, one should pay even more attention to the foods that they are eating. Mother Bear nodded after she learned about Little Bear's condition. As soon as Mother Bear arrived home, she asked Dr. Micro for suggestions on helping manage a better diet.

After she heard about what had happened to Little Bear, Dr. Micro dropped a sweat, it's a good thing that Little Bear's condition were stable then. She explained to Mother Bear, "In such hot weather, it's best not to eat even 'hotter' foods like fried chicken."

Fried chicken is one of the classic

美食城堡的安全保卫战
A War to Protect Food Safety

食城堡的居民早已听闻过：一只炸鸡腿的危害等同于60支香烟。虽然此说法并无明确考证过，但是炸鸡带给人体的危害不容小觑。

尽管鸡肉的蛋白质含量丰富，但本身就缺乏膳食纤维与其他水溶性维生素。当鸡块经高温加热后，其营养成分会遭到不同程度的破坏，且随着油温的升高和煎炸时间的延长，营养成分被破坏的程度就更加明显。因此一通煎炸之后，鸡块的营养已所剩无几。如果不注意其他食物如蔬菜、牛奶的补充，长此以往必然会造成青少年营养失衡，影响生长发育。

为了使炸鸡达到松香脆软的口感，商家通常会选用棕榈油等饱

fried foods, and the citizens of Castle City have learned all about it already: a single fried drumstick equates to the damage of smoking 60 cigars. Even though this saying was not proved scientifically, friend chicken is still extremely bad for your health, and it cannot be overlooked.

Although chicken meat is rich in protein, it lacks healthy fibers and water-soluble vitamins. The chicken meat's nutrition will be damaged to different degrees when its heated with high temperature, as the temperature of the oil increases, the degree of damage will increase as well. After being fried in oil, the chicken meat has almost zero nutritional value left inside.

In order to achieve the soft and crispy taste of fried chicken,

美食城堡的安全保卫战
A War to Protect Food Safety

和脂肪酸含量高的油，并且会在鸡块表面裹上一层面糊。这样一来，炸鸡表层的面糊将会使人摄入更多的油脂。高热量、高油脂而又难消化的炸鸡，会给人体消化系统带来巨大的负荷，引起腹部饱胀等不适症状。其中的饱和脂肪，还会使人体胆固醇升高，诱发高血压、糖尿病等心血管疾病，使成人病"年轻化"。除此之外，商家为节约成本，煎炸鸡块的油通常会被反复使用。食用油经反复高温加热，分子结构发生变化，其产生的有害物质都可能成为潜在的癌症诱发源。

值得警觉的是，炸鸡因口感独

businesses usually use palm oil and other saturated fatty acids that are high in oil concentration, and then the chicken is coated with a layer of batter. As a result, the batter on the surface of the fried chicken will absorb a lot of oil. Fried chicken is high in calorie and fat, it is also very hard to digest. The saturated fat in the fried chicken also raises the body's cholesterol level, it can induce hypertension, diabetes and other cardiovascular diseases. In addition, businesses often use the frying oil repeatedly in order to save cost. The molecular structure of the cooking oil will change after it has been heated repeatedly, and harmful substances can be created through the process that may become potential cancer inducing sources.

One thing to keep in mind is

A War to Protect Food Safety

特具有一定的"成瘾性"。这就会使人们难以控制进食量，导致人体内分泌系统发生变化，从而引发孩子肥胖、智力减退、免疫力下降，甚至性早熟等健康发育问题。

听到这，熊妈妈不禁懊悔不已，胖胖熊之前吃的炸鸡实在太多啦。但幸好有小微博士及时支招，熊妈妈暗自下定决心，以后一定要严格管理胖胖熊的饮食。

【小微博士有话说】

1. 炸鸡多吃无益。但毕竟"吃"的目的不仅仅是吃本身，有时更是为了满足某种精神上的享受。因此，小微博士建议，若实在抵挡不

the unique taste of fried chicken has addictiveness. This addictiveness makes it hard to control one's food intake, causing changes in the body's endocrine system, which leads to child obesity, mental retardation, decreased immunity, and even premature puberty and other health problems.

After listening to that, Mother Bear was extremely regretful about allowing Little Bear to eat so much fried chicken. It's a good thing that Dr. Micro gave good suggestions, so Mother Bear was determined to strictly control Little Bear's daily diet.

[Dr. Micro's notes and tips]

1. Although it's bad for you to eat a lot of fried chicken, "eating" is not only for filling your stomach, it's also for satisfying your taste buds

美食城堡的安全保卫战
A War to Protect Food Safety

图 21　炸鸡与啤酒

Figure 21　Fried chicken and beer

住炸鸡的诱惑，可选择自己制作享用。一来可以保证鸡肉的新鲜度，二来也可对食用油的安全性放心。但在食用炸鸡时，最好搭配富含维生素和抗氧化剂的果蔬，如猕猴桃、芋头、油菜、豇豆等。

and bringing a sense of enjoyment. Therefore, Dr. Micro suggests that if you really cannot resist the temptation of fried chicken, then you can choose to make your own. This way, you can make sure the chicken you pick is fresh and the frying oil is safe. But when you're eating fried chicken, it is best to match it with fruits, vegetables and antioxidants such as kiwifruit, taro, lettuce, cowpeas and so on.

A War to Protect Food Safety

2. 至于网红"星星餐"——炸鸡配啤酒,小微博士极其反对。首先,两者都属于高热量食物,啤酒向来有"液体面包"之称,而炸鸡的热量更是惊人。其次,冷啤酒和热油炸食品的搭配,更容易刺激肠胃,损伤胃黏膜,导致消化不良,引起急性肠胃炎、消化道溃疡等肠道问题。最后,这样的组合很可能会引起人体尿酸增高,导致高尿酸血症,甚至引发痛风。

3. 高温天气,人体肠道也会变得敏感。小微博士建议饮食要以清淡为主,少吃燥热的油炸食品及强刺激性的冷藏酒水。

2. Dr. Micro is strongly against the korean drama's popular "star fried chicken meal" — fried chicken and beer. First of all, both of them are high calorie foods; beer is known as "liquid bread", and the number of calories in fried chicken is astonishing. Cold beer on the other hand when matched with hot fried chicken can irritate your stomach, damaging gastric mucosa and resulting in food poisoning, acute gastroenteritis, stomach ulcers and other intestinal problems. Such combinations can cause uric acid levels in your body to increase, causing hyperuricemia and even arthritis.

3. The human intestines can be very sensitive in hot weather. Therefore, Dr. Micro suggests to eat light foods and avoid fried foods and cold beverages that can cause irritation to the body.